D0006940

Invention

Invention

The Care and Feeding of Ideas

Norbert Wiener

with an introduction by Steve Joshua Heims

The MIT Press
Cambridge, Massachusetts
London, England

This book was set in Bembo by The MIT Press and was printed and bound in the United States of America.

Library of Congress Cataloging-in-Publication Data

Wiener, Norbert, 1894–1964.
Invention : the care and feeding of ideas / Norbert Wiener ; with an introduction by Steve Joshua Heims
p. cm.
Includes index.
ISBN 0-262-23167-0
1. Inventions—History. 2. Inventions—Social aspects—United States—History. I. Title.
T15.W65 1993
303.48'3—dc20 92-35656
CIP

To the Massachusetts Institute of Technology, a home for the creative intellect

Contents

Introduction

Steve Joshua Heims

The subtitle of this book—written in the 1950s but published here for the first time—might lead an unsuspecting reader to guess that its author was an industrial manager of some sort. He was not. Norbert Wiener (1894–1964) was, in fact, a passionate intellectual and a highly original innovator in a variety of fields. His most remarkable inventions were in the domain of mathematics. Of course, we usually think of abstract mathematical ideas as part of pure science and not as inventions. They are, after all, not patentable, and once they are announced and explained, they are available to anyone who cares to use them. They also lack the palpable, tactile, visible concreteness of machines. Wiener was no ordinary mathematician, however; from his boyhood he was fascinated by machines, and he enjoyed studying their details. He linked his own mathematics to engineering, liked to work closely with engineers, and provided basic ideas for the design of all sorts of "inventions" that fit our common image.

A major example of Wiener's inventiveness was the statistical theory of communication, an outgrowth of his work during World War II. This was a general mathematical theory, but it would henceforth inform the design of every kind of communication system, from telephone networks to satellite relays and office computer networks. The problem he faced was this: Electronic communications inevitably carry not only the information we want to send but also unintended and unwanted noise. Separating the information from the noise was often a major problem. Wiener's characteristic approach was to develop a mathematical theory of how one might best filter out noise in a broad class of situations. This theory immediately led to great improvements in radar observations of aircraft and was also successfully applied in the design of noise filters for radios, telephones, and numerous other devices in common use.

As another example, in the 1920s, long before modern computers existed to help solve mathematical problems, Wiener made a concrete computational invention. He devised a method for evaluating a general class of integrals by passing a beam of light through carefully designed screens and measuring the intensity of the emerging beam. This was in effect an analog computer; it became known as the "Wiener integraph" and was later refined by others to become the "cinema integraph." By 1940, when Wiener was a mathematical consultant for "war-preparedness," he recommended as the best direction for computer development a new kind of machine: It should be digital (rather than analog), it should use binary numbers, it should be electronic, its logical structure should have the architecture of a Turing machine, and it should use magnetic tape for data storage. Unfortunately, his prescient memorandum sat ignored in an administrator's office for years; but the same ideas were independently discovered by others, and within a few years the modern high-speed digital computer was born.

A different invention came out of Wiener's application of his 1920s mathematical work on Fourier series to the analysis of electrical circuits. He realized that if one could analyze, one could also synthesize and thus know in advance from the theory how a network would perform. Wiener had some ideas about the design of a network that could be easily adjusted to give (within limits) whatever performance characteristics one wanted for a particular application. His student and later colleague Y. W. Lee worked on the engineering side of the problem until it became a well-designed, practically useful network. In the 1930s the Lee-Wiener network was patented and seemed to have commercial possibilities. The inventors' experience with the patents was unsatisfactory, however: They sold their rights to the Bell Telephone Company, which sat on the invention and also prevented anyone else from using it for seventeen years, until the patents had expired.

After World War II, Wiener became interested in the problem of prosthesis—particularly how to design a device to replace an amputated limb, including its tactile sensibility, and how to create a mechanism to replace hearing for a deaf person. A 1985 reviewer of Wiener's work on this topic wrote that "although none of the specific prosthesis projects he had suggested or in which he was involved had borne fruit as of his death (1964), the criteria he set down are still valid." For example, the successful "Boston Arm," which eventually came out of a project Wiener had initiated, exemplified precisely the principles he had anticipated would work.

These examples suffice to show Wiener's characteristic contribution to the invention of new technologies. He had the capacity to understand how things worked or might work on a very deep level, but he usually left it to others to translate, and sometimes modify, his conceptions to create useful hardware. As our first two examples suggest, he was an instigator, very far-sighted in matters of new technological developments, helping to launch the so-called Second Industrial Revolution, which has been marked by the development and widespread application of new technologies of communication and computation.

Wiener's experience with the Lee-Wiener network and his suggestions for digital computers heightened his sensitivity to the unfairness with which inventors can be treated, especially if they are not wise in the ways of the world. He was remarkable for being equally concerned with original thinking in mathematics and engineering and with understanding the social and philosophical impact of his own and his inventive colleagues' handiwork. It was this combination that had most intrigued me when as a college student I first read Wiener's *Cybernetics, or Control and Communication in the Animal and the Machine* (1948). It was totally unprecedented to combine in a single, relatively short book highly interesting, and often original, scientific and technical work with an intense concern for the likely social and political applications of that work.

Notwithstanding its use of advanced mathematics and a style that was untidy and discursive, *Cybernetics* unexpectedly became a best-seller. Wiener came to be widely known and a popular lecturer. As a result, he decided to complement his scientific work with books addressed to the general public. His next book, again discursive but without the mathematics, describes the ideas of *Cybernetics* specifically for the layperson. It was called *The Human Use of Human Beings* and was first published in 1950 by Houghton Mifflin. In a second revised edition, published by Doubleday in 1954, the writing and organization are a bit tighter and more orderly, but the pungency and bluntness of some of the comments in the first edition are lost. Jason Epstein of Doubleday had been the editor.

Wiener finished the first volume of his autobiography, *Ex-Prodigy*, in 1952. By 1954 he was working on a second volume, *I Am a Mathematician*, with advice from Epstein. Since Wiener tended to be uninhibited in expressing his opinions about colleagues, Epstein was at pains to tone him down to avoid any possibility of libel. The book would appear in print in 1956.

Meanwhile Epstein suggested that Wiener write a book on the philosophy of invention for a broad general readership that Doubleday would publish as an inexpensive paperback. Wiener accepted a $500 advance and wrote, or rather dictated, the first draft for the book in April 1954. The present manuscript, found among Wiener's papers in the Institute Archives of the MIT Libraries (which are also the source for the correspondence quoted here), is the latest existing revision, dated June 1954.

The correspondence with Epstein in this period deals practically only with Wiener's autobiographical volume, and the "Invention" manuscript was apparently relegated to a back burner. Later in 1954, after a trip to India, the prolific Wiener was engaged with the mathematical analysis of brain waves, a reformulation of quantum theory, and the sensory prosthesis project. He was also starting to contemplate a more literary work. He had for some time been thinking about the theatrical or novelistic possibilities of a dramatic tale in the history of invention in which Oliver Heaviside, a pioneer in the theory of telephonic communication, would be the hero while the American Telephone and Telegraph Company and the engineer Michael Pupin of Columbia University would be the villains. In 1941 Wiener had sent an outline of the plot and the major characters to Orson Welles, inviting him to use it as a basis for a film, but nothing ever came of this offer.

Correspondence between Wiener and Epstein in August 1957 sheds light on why *Invention: The Care and Feeding of Ideas* was never published. Wiener was becoming more and more fascinated by the Heaviside story (which is also told in chapter 6 of *Invention*). He wrote to Epstein on August 2, 1957: "I have been proceeding with my work on the scenario of the Heaviside story. I have decided that the best way of handling it is to write it completely as a novel. The work is going smoothly. . . . The story is really a treatment in fictional form of my ideas on invention in the modern world. I am not too enthusiastic about carrying through my plans for a purely expository book on invention. I therefore suggest that you take this manuscript in lieu of such a book, and that I accept the moneys you have advanced me as a part of the advance on the publication of this book."

Epstein replied that he was eager to see the manuscript for the novel, but "I have again read the manuscript on invention . . . , and except for several instances where you don't explain clearly to the reader the technical problems that arise, the book is extremely good and should, after some work, be published. Is there any reason why it can't be done as well as the novel?"

Notwithstanding Epstein's high opinion of the manuscript, Wiener had lost interest in it. He returned the $500 advance to Doubleday with a note: "I have decided to abandon the proposed book on invention of which you have a first draft ms." Epstein had meanwhile moved from Doubleday to Random House, which would publish Wiener's novel, *The Tempter*, in 1959. There is no further mention of the invention book. This was a loss for the reading public, since Wiener was a gifted expository writer but his talent as a novelist was not great.

In reading the manuscript on inventions in the context of Wiener's other personal and public writings, I am struck by its exceptionally hopeful tone, which contrasts with the more pessimistic outlook he expressed elsewhere. It is just possible that when he went back to the 1954 draft in 1957, he found its implicit optimism unrealistic. As with all of his ideas, he would be inclined to worry that his suggestions for fostering new inventions would be taken up most readily by people whose purposes he abhorred. Not wanting to fuss with the manuscript further, he was content to relegate it to his files.

The manuscript shows primarily one side of the duality of Wiener's attitude toward inventions. He refers to the "desperate necessity" for technological inventions for the basic survival of life and civilization, and he insists on the necessity for scientist/inventors to be totally dedicated to their work and for the larger community to have faith that the innovators' ideas and gadgets will redound to the public benefit. In all it is a hopeful advocacy, but one requiring faith and commitment, as well as an unending stream of new fundamental ideas and new kinds of machinery.

Wiener cared about the long-term future of humanity, and these new ideas were to be applied to redress long-term global problems—Wiener refers to malnutrition, lack of clean water, exhaustion of natural resources, poisoning of the environment, etc.—but here is the rub. The extant political and economic systems tend to respond to short-term crises but do not support the kinds of ideas and activities that are likely to be crucial for the more distant future. In the United States, for example, political leaders elected for two, four, or six years at a time tend to be concerned only with the short-term effects of their policies. Wiener's examination of the economic system of private enterprise led him to conclude that businesses, too, "are relatively short-time undertakings, and are not able by their very nature to pay much attention to the long-time secular interests of the human race." Elsewhere Wiener had sounded generally discouraged about the prospect of innovations being applied in a primarily benign way, and he was constantly enjoining scientists and engineers to act with high morality in this regard.

Of the set of ideas and techniques he called "cybernetics" (in part his own ideas) and their applications he wrote that they had "unbounded possibilities for good and for evil." But he also stressed that inevitably the work of scientists and engineers comes into the hands of the powers-that-be in society. Wiener also expressed pessimism concerning the mode of operation of large for-profit corporations, totalitarian governments, and especially the military sectors of society. He argued in *Cybernetics* that these groups are all likely to use inventions in distinctly harmful ways. He expected that in our world of oppressive governments and profit-oriented corporations, many inventions would be used primarily to hasten the accumulation of power—which, according to Wiener, "is always concentrated, by its very condition of existence, in the hands of the most unscrupulous"—and would thus become detrimental to the welfare of the wider public.

In his own work Wiener found a way through these conflicting moods and was able to honor both his love of innovation and his skepticism. His hopeful side allowed Wiener to continue working on fundamental mathematical and engineering ideas. His pessimistic side led him to warn the public about the possible misuse of inventions and inspired him after the end of the war to participate only in what he judged to be the most benign applications.

The first half of the present book is an engagingly written narrative of the history of discovery and invention, with many interesting sidetrips. Since writing history always entails some interpretation, and Wiener not only had a cross-disciplinary grasp of the history of invention but also knew the process of invention from the inside, it is of considerable interest to see which topics he selects for emphasis, which historical circumstances he sees as fostering innovation, and what connections he makes. Characteristically, he begins with mathematical ideas, the historical origins of arithmetic and geometry, and goes on from there to more advanced concepts in mathematics and mathematical physics. He argues that since "mathematics allows us to state the essential" without distraction by the inessential, and since its abstractness transcends any particular field of application, it can be "a powerful organ of invention and discovery." "Ideas," however, are only one element in useful invention.

Another element in invention, Wiener asserts, is the availability of techniques and materials. To explore their role, he takes as his first example the inventions described in Leonardo da Vinci's notebooks, many of which could be adequately realized only several centuries later, once techniques for working and machining metals and the use of lubricants became more sophisticated. His second example—optical instruments, microscopes, and telescopes—illustrates the confluence of techniques of lens-grinding, polishing, and instrument-construction with the mathematics of optical instruments. Wiener goes on to discuss clocks, sailing ships, methods for determining latitude and longitude, and at somewhat greater length the history of steam-engines. This leads him to consider some of the many inventions that originated in China but eventually made their way to the West, and he comments on the superiority of Chinese to European civilization up to the time of the Renaissance. After describing the Chinese invention of gunpowder and the development of cannons, Wiener closes the chapter with comments on nuclear weapons.

One aspect of social climate that favors invention is when scientists (or philosophers) and craftsman (or engineers) can communicate directly with each other and are not separated by social class. In chapter 5 Wiener traces this theme in large strokes from ancient Greece in the fourth century B.C. through the Middle Ages in Europe, the political revolution in France, and the Industrial Revolution in England, ending with discussion of such men as Benjamin Franklin, Count Rumford, Faraday, Maxwell, and Lord Kelvin. The next chapter describes Thomas Edison's social innovation, the industrial scientific laboratory, as well as the history of the telephone and the development of the electrical communication industry. Here Wiener tells the story of the transatlantic cable and the minimizing of distortions in telephone transmission and describes how the management of an American company connived with a Columbia University professor to deprive the eccentric inventor Oliver Heaviside of his just due.

Wiener's survey of the history of technology is remarkable for the multitude of technically detailed examples he describes in simple language. It is the rich texture of concrete details, at least as much as the general statements, that makes the book enlightening reading. The style is easy, nearly conversational, and full of anecdotes—in all, a brief, pleasant, and brilliant introduction to the history of invention, as suitable for the general reader or student today as it was when written.

Wiener was not Eurocentric, as many previous historians of science and technology had been. Here, for example, he touches on the question of why China was for so long far more advanced than Europe in technology and invention, and he suggests that the explanation has to do with the relatively high standing of scholars and craftsmen and the low standing of soldiers, merchants, and businessmen in Confucian ideology. This interesting historical problem has since been explored systematically and in depth by Joseph Needham, whose essays—for example, those in *The Grand Titration*—are a useful supplement to Wiener's discussion.

Similarly, Wiener's discussion of the social conditions of the French Revolution and the Industrial Revolution might well be supplemented by some general social history of that era, such as Eric Hobsbawm's *The Age of Revolution 1789–1848*. To get a sense of the extent of the arbitrariness of emphasis and interpretation in much of the history of technology, the interested reader might also take a look at Arnold Pacey's very readable *The Maze of Ingenuity*.

In the second half of the book Wiener addresses issues and problems related to inventions and their connection to social and economic patterns in the United States. Even though he was writing during the cold war, a decade after World War II, he was unaffected by cold war fears, militarism, or secrecy. It was anticipated at that time, surprisingly accurately, that the cold war would continue until the early 1990s, and Wiener cared especially about the longer-term future of humanity. "Our long-time chief antagonist," he wrote, "is to be sought for among the continuing threats of hunger, thirst, ignorance, overpopulation, and perhaps the new dangers of the poisoning of the world in which we live." Since Wiener did not appraise the situation for inventions within the framework of short-term considerations such as the cold war or financial profit-and-loss, but rather within the framework of the long-term needs of humanity, his comments continue to be relevant to the post–cold war present.

Chapters 7 and 8 describe Wiener's thoughts about the social environment of invention during the postwar period, which he calls the era of "megabuck science." He was unhappy that the large research laboratory, especially the industrial laboratory, with its well-defined missions and the high degree of specialization among its employees, was becoming the norm, and he expresses a fear that the independent free-wheeling scientist and inventor may become a rarity. He laments the increased prominence of "scientific adventurers" on the make, with no true devotion to science for its own sake. He is further distressed by the increasing prevalence of secrecy in research and in technical developments, and he argues that there is much harm in this method of operation. In the 1990s, with money-making still an all too prominent social consideration, it is refreshing (despite an inevitable wince at the sexist language of the times) to read Wiener's statement that "the real scientist of the first rank is by the nature of his own activity too busy a man to care much either for money or the ordinary signs of prosperity."

While Wiener's concern about the directions the organizational patterns of science and invention were taking is evident, his critique of these directions is thoughtful, and the points he makes are always interesting and often persuasive. Interwoven with Wiener's critique is a review of the scientific history of the atomic bomb and the unpromising efforts to design machines that automatically translate one language to another without human intervention. In contrasting mission-oriented research with a more free-wheeling style, he is led to the notion of the "inverse problem of invention": "At many stages we possess new constructible tools or new intellectual tools which obviously are bound to increase our powers considerably in some direction or other. The question is, in what direction? It is . . . truly a work of invention or discovery to find out what we are able to accomplish by the use of these new tools." He illustrates his point with histories of the electric motor and the vacuum tube. Today he might have chosen the personal computer. In chapter 10, Wiener describes some unsatisfactory features, and resulting inequities, in the U.S. patent system, and he concludes with an argument that society benefits from and should figure out a way to reward independent "free-lance" scientists.

Invention thus offers an excellent guided tour of the history of invention, together with many thoughts and observations about the conditions that favor the emergence of new ideas. Because it describes the care and feeding of scientifically or technically innovative minds, it will be of particular interest to anyone concerned about issues of education and technical competitiveness. Scientists, inventors, and engineers, as well as students just embarking on such careers, are also likely to find the book of value as they reflect on their own needs in relation to present or anticipated future working conditions.

While he acknowledges that the raison d'être for innovation lies in the requirements for human living, it must be noted that Wiener does not focus attention on the user's view of inventions and the design of technologies, and his philosophy of invention is therefore not a general philosophy of technology. From the perspective of users or consumers, the book has little to say beyond the axiomatic assertion of the necessity of invention for collective survival and the general observation that neither cold war activities nor market mechanisms are geared to the long-term public interest.

These general observations apply with equal force today. Markets and their needs again dominate worldwide economics. Today's counterpart to cold war activities is the continued diversion of limited resources to further weapons development and military projects of all sorts, including wars. Every day we are reminded of the progressive deterioration of our environment and the depletion of our vital human necessities. In the 1970s Nicholas Georgescu-Roegen, Hazel Henderson, André Gorz, and others began to analyze and write hopefully about what we need to do to survive. Today, while a few research groups and some other independent organizations approximate islands where a humane global long-term perspective is the norm, they are not a powerful element in society, and the political and economic mechanisms available to implement their recommendations are not very effective. The unavoidable conclusion, reinforced by the events of the last four decades, is that a great deal of social, political, and economic innovation and change is still required to ensure that existing and future technologies truly serve all segments of the world population. These political issues are controversial, but our need to address them now is even more urgent than it was in 1954.

Preface

This book has been solicited from me by Mr. Jason Epstein of Doubleday and Company and the Anchor Book Series, as a result of several conversations we have held together without, in the beginning, any particular idea that I should publish this material. I am deeply grateful to Mr. Epstein because I think that an inexpensive paperback series like the Anchor Books is exactly the right place to publish such material.

If this book has any value at all, it is to call people's attention to what is happening in the intellectual world and the world of invention, and to call their attention to these things fundamentally for the purpose of getting them to take a definite attitude towards what I consider to be the unfortunate trends of the present day.

An appeal of this sort is of little use if it is confined to a narrow circle, whether of professional intellectuals or of those who feel themselves in a position to buy $4.00 or $5.00 books. Thus the form in which this book appears is an intrinsic part of whatever value it may have.

I have discussed this book with several of my friends, and in particular with Professor Karl Deutsch and Mr. W. D. Stahlman of our Department of Humanities. Both of these gentlemen have volunteered me many detailed and positive criticisms, a large part of which I have incorporated into the text. I hereby express my debt of gratitude.

My habitual mode of working is by dictation to a secretary, not, I may say, by dictating into a dictaphone. In this sort of work the active and implied criticism of the secretary is of great value to me. I wish to thank Barbara Beaumont Cole for her participative and sensitive help in the writing of this book.

Cambridge, Massachusetts
June 1954

Invention

The Need and Conditions for Invention

1

The present book is in one sense the result of the reflection of thirty-five years spent at the Massachusetts Institute of Technology in intimate connection with engineering, scientific, and economic developments. On the other hand, it represents a specific response to a request of Mr. Jason Epstein for a book on invention for the Anchor series of books published by Doubleday. The subject has long been an exciting one for me, and I have welcomed the opportunity to write a book on invention, with free license to put in what I have wanted and to leave out what I have wanted.

We are living in an age that differs from all previous ones by the fact that the invention of new machinery and, in general, of new means of controlling our environment, is no longer a sporadic phenomenon, but has become an understood process to which we resort, not merely to improve the scale of living and the amenities of life, but from a desperate necessity to render any life whatever, and certainly any civilized life, possible in the future. For many decades people have been talking of the approaching exhaustion of certain of our natural resources. We have had the rather unformed idea that when the time should come that our shortages of raw material should reach an acute point, something in the way of new technical developments might be expected to come and save us.

Until recently, however, this approaching exhaustion of our essential resources was put, by most writers, as a new state of affairs that we might have to face in centuries or millennia or, at worst, in half a century, but not within the effective life of any large fraction of the population now alive. The ever-increasing growth of technique, and the particularly accelerated growth due to a couple of great wars and a prolonged period of military tension, have made many of these shortages matters of reasonably present concern, to which we must devote at least a considerable part of our planning potential at this day and moment.

Even the most common and most inexhaustible metals, such as iron, are only maintained in our stockpiles by a continued development of techniques that allow us to supplement such standard sources of supply as the depleted Mesabi range by newly discovered supplies in remote quarters of the globe, and by techniques that permit us to break the energy barriers which hold the metals confined in inferior ones. By these energy barriers, I mean both the mechanical problems of handling hard, bulky, and unfruitful materials, and the problems of smelting at higher and higher temperatures. Similarly and far more acutely, copper, lead, and particularly tin have almost attained the status of precious metals; while dozens of metals which have traditionally possessed this status have now ceased to be laboratory curiosities and have become part of the basic materials of industry.

Notwithstanding the art of irrigation, which is as old as any high form of culture, we have come to consider water as nearly a free good over a large part of the civilized world; and what is more, as a free good that can endow us with a considerable fraction of the energy needed for running our industries. New supplies of water power are no longer in sight in quantities comparable with the growth of our needs for power, and the more commonplace uses of water for supplying human thirst, for cooling thermal engines, and as a raw material in its own right to be exploited through industrial processes, are meeting the almost worldwide emergency of a falling water table in the soil. Thus the very survival of cities like Los Angeles, in a region of arid climate, and yet on the edge of that tantalizingly unusable mass of water, the sea, is conditioned by our actual progress in the direction of economically practicable freshening sea water, and by our far less modest hopes for the future of this as yet imperfectly developed craft.

Human hunger offers an even greater challenge than human thirst. Not only have the economic and social boundaries of what was the Western world of a century ago been so widened that they now take in regions of dense populations on the perpetual edge of famine, but we have also come to the awareness that this problem has no limits unless we put some control on the all-engulfing force of human fecundity. But even if we stem this tide of reproduction in which, like the locusts or the lemmings, we eat our individual selves into what may be a racial grave, no one is satisfied with the present level of undernutrition on which a good half of the human race lives. The inefficient process of converting solar energy into grass and grass into meat, upon which a large part of the countries of higher standards of living depends, is clearly too wasteful even for the moderately immediate future. Improved methods of animal husbandry, of agriculture, and of fisheries can and must stave off the crisis for a matter of years or decades, but it is no accident that we are now forced to revise our use of solar photosynthesis back to its first principles and to explore the possibilities of the unicellular algae as producers of fat, carbohydrates, and protein.

Thus we live only by the grace of invention: not merely by such invention as has already been made, but by our hope of new and as yet nonexisting inventions for the future. Since we are committed to a life in which the process of invention is not considered any more merely as a source of human capital but as a part of human income, we should contemplate very carefully the nature of this income process, and the degree of regularity or irregularity with which we may count upon it for the future.

Now, the history of inventions, the psychology of inventions, and, in general, the principles that cover the occasions in time and in the development of scientific and human needs when inventions are likely to be made remain extremely obscure. For one thing, there is a strong fortuitous element in invention, and the mere need for a discovery, while it may direct effort into that region of work where the discovery may be expected to arise, and while it in the long run will favor such a discovery, gives no assurance whatever that the discovery will be made within a certain limited amount of time.

I have been much impressed by the history of invention and discovery as a battleground of a most intense conflict between two notions of history. In one of these notions, which has been the favorite point of departure of most historians down till the end of the last century, history is largely a theater in which kings and statesmen and generals and great names play the leading parts. On the other hand, from Marx and Engels on, we have been taught to regard history as essentially an interplay of economic and great social forces, in which the individual means little more than a somewhat accidental embodiment of these forces. The actors are secondary to the Greek chorus.

Kipling has emphasized the difference between these points of view in a dramatic way, in discussing the difference between English and American patriotism. According to him, and I believe that here he is quite just, English patriotism centers about the king, and American patriotism about the flag. "There is too much Romeo and too little balcony about our [the English] National anthem," he says. "With the American article it is all balcony" (*From Sea to Sea*, Vol. 2, No. 36). The part that the king plays is that of Romeo, and the part that the flag plays is essentially that of a drapery for the balcony. In a historical treatment of science and invention, what are the relative roles of Romeo and of the balcony?

The personal, Romeo, point of view is brought out in Pope's verse "Intended for Sir Isaac Newton":

Nature and Nature's laws lay hid in night;
God said, *Let Newton be!* and all was light.

The economic theory of history is all balcony and no Romeo. This theory, in addition to being an intrinsic part of Soviet ideology, has spread widely in the West, both in practice and in theory, among social elements which would be shocked at sharing anything with Marx and Communism. It is in accordance with this view of science, scholarship, and invention that the emphasis has shifted from the individual learned man working in a university laboratory or in his own home as a free-lance, to the great aggregate of scholars working together in an industrial or government laboratory where each contributes his little bit to the vast job of assembly.

On the intellectual level, this development is a precise counterpart of the change from the craftsman who constructs an entire carriage, to the piece worker in an assembly line who spends his whole life in screwing in one particular bolt. Marx was well aware of the change on the craftsman-industrial level, and he does not seem at any time to have been averse to a view accepting a similar anonymity of the workman and socially controlled effort on the level of scientific discovery. At any rate, whatever the ideological differences between Russia and the United States, the actual trend towards large-scale investigations involving divided and even comminuted responsibility turns out to be the same in the two countries. I think it matters little whether the agencies of this subdivision of thought and responsibility be commercial laboratories or community- or state-organized ones.

Both sorts of institutions find themselves allotted a budget of dollars of effort strictly assigned to particular tasks, and in both the scholar-workman is bound to a perpetual subordination to a prearranged order of things. Notwithstanding this, both in Russia and in the United States, some and even a great deal of attention is paid to science on a more universal and abstract level and operating with a greater freedom. The site of this activity may be a university or an academy; but in both countries, in the eye of the community at large as well as in the eye of the high-level administrator, the free-lance scientist without a range of responsibility delegated in advance, while he may be regarded as necessary, has become the stepchild of scientific development instead of the favored son. In America and, I am quite sure, in Russia as well, people like James B. Conant realize the danger that the prodigal stepson may be driven away to starve with the swine, but there can be no doubt that this recognition of his necessity is a bit grudging and forced in both places.

Thus one of the purposes of the present book is to make a proper assessment of the individual element in invention and discovery and of the cultural element. This problem is rendered more complicated by the fact that there are several stages of invention in which the relative worths of the individual and the environment are quite different. Before we even start this assessment, however, we must have notions somewhat clearer than our naive notions of what influence and causality mean, for only within such a frame does the question of the balance between the individual and the environment have significance.

In order to assess influence and causality, we must face a world in which causality is in some way localized. If the whole world of the past causes the whole world of the future, and nothing more specific can be said, then causality is too general to be a useful notion for analysis; but if we are to consider causality in more than this vague sense, and to contemplate varying quantities of causality, we must consider a world in which imperfect causality can be measured. The tight world of Newtonian physics never mentions causality at all, and this is because the technique of assessing causality is to relax one or more factors in the world and to see how this relaxation in the past affects the present and the future.

The situation is very much like that which we meet in a truss bridge. To determine the strains that the different parts are carrying, we place a weight somewhere on the bridge and measure the deflection of element after element. If there is no deflection, and if the bridge is of the sort which we ordinarily call overdetermined, which means essentially that it does not collapse at the removal of any one element, then for all we know some one element may be carrying vastly more strain than what it is designed to carry, and the bridge is unsafe.

This is not merely a theoretical difficulty of a carping intellectual but an actual cause for the failure of bridges. The welded bridge can only come into equilibrium by giving short of failing, and the welded bridge of a material so rigid that it does not give perceptibly has often collapsed without an ascertainable reason, through the unobservable bad distribution of its internal stresses.

In a similar way, for a system to show any effective causality, it must be possible to consider how this system would have behaved if it had been built up in a slightly different way. For example, if we wish to compare the relative importance of Newton and of Edison for the history of invention, we must be able to consider in a reasonable manner both what that history would have been like without Newton and what that history would have been like without Edison.

The thesis that I shall try to support in this book is that in the process of invention at least four important moments arise, some of them early and some of them late. Before any new idea can arise in theory or in practice, some person or persons must have introduced it in their own minds, and this change must have come to be preserved in accessible records, thereby causing a change in the intellectual climate. At this stage, the effectiveness of the individual is enormous. The absence of the proper original mind, even though it might not have excluded a certain element of progress in the distant future, may well delay it for fifty years or a century.

The second element favoring invention is the existence of proper materials or techniques. These are not, in truth, a part of the original idea, but may be necessary for its effective execution. Later I shall give several examples of these apparently extraneous changes in materials or techniques.

Once a new idea is at hand and has been recorded in writings or in generally accepted concepts that are available to the craftsmen in the field in terms of existing materials and techniques, the particular date at which the effectively usable invention is made is highly indeterminate. Under these circumstances, the same invention is more likely than not to be made independently in many places. This is the stage of invention which belongs rather to the balcony than to Romeo, and in which an economic theory of scientific and industrial development is particularly applicable.

However, before a new technique can pass from the intellectual to the artisan, these two very different types of men must have an adequate means of communication with each other within the social system in which they live. To paraphrase Plato, the artisans must become philosophers, or the philosopher, artisans. In the very Platonic period of Periclean Greece, the communication between artisan and philosopher was perhaps at the lowest level it had attained or was to attain in civilized times. The old Cretan respect for the ingenuity of Daedalus had been decreasing for many centuries. In the period of the great days of Athens, Pericles repealed an old law permitting specially skilled craftsmen to become citizens. In this he acted in the capacity of our Senator Patrick McCarran.

After the age of Pericles, the Greek city-state broke down, and the culture that had been confined to a narrow literate ruling class spread by a spiritual osmosis to the new cities of Alexandria or Syracuse, where Greeks lived side by side with foreign races such as the Egyptians or Phoenicians and, above all, where new and humbler classes came to have access to the treasures of Greek culture. Thus the Hellenistic artisan was enabled to partake in the intellectual creations of the Athenian gentleman.

After these three stages of intellectual climate, technical climate, and social climate, there comes a further stage in which invention depends upon economic climate. Before inventions are made available to mankind at large, there must be a way to promote them. Under most conditions this means that there must be a process by means of which some individual or class can make a living out of promoting inventions. If the inevitable risks of a technical change are centered too closely about those people who originally make the change, and there are no countervailing means to protect these entrepreneurs, no one will dare take the risk. Under these circumstances (which indeed were the circumstances of Alexandria and largely of Renaissance Italy) a new idea or new technique will become a will-o'-the-wisp, always just out of reach of the times.

I shall devote the rest of this book to a consecutive study of these four stages of invention, to the discussion of certain specific periods, and the balance between these, and finally to the contemporary history of invention and its fate as seen through a glass, darkly.

The Intellectual Climate and Invention

2

I have already spoken about the intellectual climate as one of the chief factors in invention. I shall devote this chapter to rendering this general remark more concrete, and to the discussion of certain particularly important eras in the changing of the intellectual climate.

One of the less predictable changes occurred among the ancient Indians and apparently occurred again among the Mayans when it was discovered that in the writing of large numbers, the use of position could be brought to reinforce the use of specific characters for specific numbers. It is this use of position, for example, which enables us to write the symbol 125 for the number one hundred and twenty-five, with the meaning one hundred corresponding to a digit 1 two places to the left, twenty corresponding to a digit 2 one place to the left, and five corresponding to a number 5 in the extreme right-hand position. This combination of numbers is read one hundred plus two tens plus five, or one hundred and twenty-five.

It was the discovery of this principle by the Indians that made arithmetic possible on paper. Ideationally considered, perhaps this is too late a time at which to place the original invention which led to digits of position. In the simplest form of the abacus, as known both to the ancient Greeks and Romans and to the Chinese, the motion of a number of beads was used to indicate a digit, and the particular wire on which these beads were suspended indicated the position of the digit. Thus we had already in very ancient times a positional notation which was not a written notation, but which was sufficiently clear and manipulable to render the abacus a splendid tool of computation, replacing paper computation for all those peoples who know the abacus but do not have a written positional notation, and indeed, for many who can do their computations alternatively on paper. This computation by the abacus, and by the ideal abacus of the positional system of notation, was discovered independently by the Mayans for the scale of twenty.

What characterizes these inventions historically as an intellectual triumph is the relatively small number of times at which the idea arose in the course of recorded history, and the tremendous discrepancy of time between its repeated discoveries. Obvious as the discovery of positional notation seems to us at the present, ages have passed in which the concept would have been useful and certainly was theoretically discoverable, but in which it was not in fact made. It was far from being at all clear that this discovery was about to be made at any of the times that some unknown thinker produced it out of empty air, and the world could have gone on for centuries more with positional notation still undiscovered, and even with no abacus. When these two closely related discoveries were made, the craft of the merchant and the artisan became much easier, and the usefulness of the new concepts is shown by the rapidity with which they spread from one country to another, until the Indian positional notation became the common possession of the countries of its origin, the Moslem world, and Europe. Yet for all its universal spread, it was not until centuries after the innovation had been carried out that people were able to look sufficiently objectively back upon it to see what a marvelous change it had been.

Just as the notation of position had to germinate in some individual spirit before it became a heritage of the human race, so the

Greek discoveries of plane and solid geometry are to be distinguished by their lack of obviousness and by the fact that it took centuries or indeed millennia before they could be fully evaluated. The idea that lines and circles could be studied in terms of a few very simple and intellectually statable properties is elementary now that we see it, but was no early necessity of thought. That the concepts of geometry were conceived by a people with no straight edge better than a stretched cord, no compass other than a string held at one end and a stylus writing in the dust at the other, and even no geometrical diagrams beyond what could be made in this dust or on a wax table, is remarkable indeed.

That even very primitive peoples, and the Greeks were certainly not primitive, should discover properties of straight lines and circles is not remarkable, although, as I have said, it is remarkable that they should have worked up these ideas in so consistent a logical scheme, a logical scheme which, in fact, was not surpassed until well along in the nineteenth century. However, that the Greek Menaechmus should have conceived of the conic sections—the ellipse, hyperbola, and parabola—as an interesting and important field of investigation is most remarkable. That a nation of potters should not be unfamiliar with the cone is easy enough to understand, as is the idea that they should have occasionally cut the cone with a straight edge and seen what the boundary of the cut looks like, but that this more or less accidental observation should have led them to what has remained for more than two thousand years one of the most powerful intellectual tools is striking in the extreme.

The Greeks after Menaechmus continued and developed his theory of conic sections, but primarily as a self-contained field of work, and it is not until the Renaissance and the time of Kepler that we have a firm suggestion that the heavenly bodies move in conic sections. The empiric laws of Kepler again remained mere accidents of observation until Newton developed the notions not only of dynamics in general but of a particular form of dynamics under the force of gravitation which was to explain at the same time the elliptical orbits of Kepler and the parabolic orbit of a cast stone. Here again we have a setting of intellectual climate which led to discoveries millennia later, and it is easy to conceive of many more millennia passing before another Menaechmus, Kepler, or Newton might have had the key idea which has led to all modern physical science.

As to Newton, we must make a rather interesting distinction between his work on the calculus on the one hand, and his work on dynamics and gravitation on the other. The first of these had been present in petto in the analytical geometry of Descartes, and many other people—Cavalieri, Fermat, Barrow, and Wallace—came so close to the idea of differentiation that it was but a matter of a short time before one could expect the differential calculus to have a status as an independent science.

As to the integral calculus, the roots of this are to be found in the really first-class change of intellectual climate which was brought about by Archimedes in ancient Syracuse. Thus Newton is entirely right and is not governed by false modesty when he says, "If I have seen further [than others] it is by standing on the shoulders of Giants." However, the Newton of gravitation and dynamics, as distinguished from the Newton of the calculus, was the greatest of giants in his own right.

These examples may suffice to show the reader what I mean by a change in intellectual climate, and also how such a change depends on one or very few men, how it might or might not have occurred within a particular region of history, and how long one can easily have to wait from one step to the next step of similar fundamental importance. These stages of discovery have a unique individuality, and are only ill subject to a statistical discussion.

For statistics we need a sufficient population of closely similar cases to be able to observe not only what is exceptional but what is regular. The gamble of the really great discoveries is not only a gamble against tremendous odds, but a gamble against unascertainable odds. It is because of the length of these odds and of the faith that must accompany the man who is willing to lay against them that there is something divine in invention at its highest level. It is not without right that Prometheus, bound on the rock of Caucasus, with the vultures devouring his liver, cries out, "Oh godly air and swift-winged winds, and the source of rivers, and the numberless laughings of the billows of the sea, and earth, mother of all, and the circle of the sun, to be seen on all sides, behold what things I, a god, suffer at the hands of the gods."

I have just spoken of probability, of chance, and of odds. The history of the mathematics of these fields is germane to the purpose of this chapter in more than one way. Not only do we need notions of probability to assess the achievements and difficulties of the task of discovery, but the theory of probability itself is the source of a great contemporary change in the intellectual climate which we can study with all the more purpose and significance because at least one act of the drama has been played before our eyes.

Let me then turn to the theory of probability and its repercussions on modern modes of thought and modern trends in invention. The early history of this field goes back to the seventeenth century, and to such great names as those of Fermat and Pascal, but it does not go back to that part of their work which they may be considered to have developed with the greatest solemnity and the greatest sense of its ultimate philosophical importance.

Quite the contrary, it goes back to the time made known to most of us by such literary works as *The Three Musketeers* and *Cyrano de Bergerac*, when the nobles of the Court were philosophical roisterers and roistering philosophers, when the friends of a great spirit like Pascal valued not only his thoughts on religion and his baring of his naked soul, but also his advice concerning the odds on a throw of the dice or a turn of the cards. This casual advice to gamblers certainly made considerable intellectual demands on the part of those that gave it, and the satisfaction of these demands unquestionably pleased their sense of method and of internal power, but it is scarcely to be supposed that even those men who were later caricatured by Swift as the academicians of Laputa could have conceived a distant future in which their idle thoughts about play might determine the investment of many lakhs of rupees by the jute planters of Bengal, or stand subservient to the construction of the hellmouth of the hydrogen bomb. Yet every step, from Fermat and Pascal to our post-Manhattan Project age and the organization of modern industry, can be documented over and over again.

After the days of these great probabilists, physics took a turn towards determinism through that work of Newton on dynamics and celestial mechanics, which we have already mentioned and celebrated; yet even in the days of the immediate successors of Newton, it became clear that if their celestial universe were in fact the wheel of a great machine, it also had aspects in which it had to be considered as a roulette wheel. To this point of view we must attach two fabulous names—the great, if flamboyant, name of Laplace; and the even greater name of Lagrange, his older and soberer contemporary.

There are places in the solar system where the manifoldness of the heavens goes far beyond our ability to grasp the phenomena one by one, through precise measurements and a closed series of observations. Most of the mass of the solar system is in the sun and in a few great planets, but between the orbits of Mars and Jupiter there lies that sandstorm of minor planets known as the system of the asteroids. It is indeed true that the individual orbits of the asteroids can be studied by the same methods as those used for the larger planets, but it is also true that the system as a whole shows certain regularities and densities which pertain to the asteroids as a sandstorm, and which are completely masked if we look at the individual grains of such a sandstorm only as we might look at a forest through nothing but the consideration of the individual trees. It was questions of the regularity of distributions of this sort that led Laplace to make a great examination of the cosmogony of the solar system. In order to make progress in this difficult field, he was driven back to a study of the theory of probability.

Another and not unrelated tool used by both Laplace and Lagrange in connection with these studies and others closely related to them was that of trigonometric series. In the study of rotating systems it is inevitable that we should pay much attention to angles and that we should soon find that certain quantities that we wish to discuss appear naturally as infinite sums of sines and cosines of the multiples of certain angles. These sums combine additively in series now known after the name of Laplace's successor Fourier. In the eighteenth century these series caused a great deal of worry to the physicists and mathematicians, for unlike the so-called power-series used by Newton and known by the names of Newton's successors Maclaurin and Taylor, these new series seem to be useful for representing phenomena with certain discontinuities or angularities, while the Taylor and Maclaurin series are by their very nature smooth in the extreme.

Fourier, who lived at the beginning of the nineteenth century, participated as a Membre de l'Institut in Napoleon's expedition to Egypt. He showed that such series, which ever since have been known by his name, did actually possess the apparently contradictory properties with which earlier mathematicians had endowed them. Nevertheless, the subject of Fourier series is not even now closed, and it was not until the work done by Borel and Lebesgue about 1900 that we can say that mathematics was in the possession of an even approximately satisfactory theory of Fourier series, a theory of a degree of generality sufficient for the everyday needs of mathematical physics at the present time.

I have mentioned this subject of trigonometric series because, on the one hand, a great many of the reasonings of Laplace and Lagrange use them in connection with questions of probability; and because, on the other, the work of Lebesgue and Borel plays an essential role in the burgeoning of probability theory within the present generation. Nevertheless, both probability theory and the really precise theory of trigonometric series remained rather in the doldrums for about half a century. It is not until Clerk Maxwell's work on the kinetic theory of gases that we see signs in probability theory of its coming into its own again as an important tool of the physicist.

Maxwell's kinetic theory of gases represents a practical physical investigation of a problem that had been waiting for the mathematicians since the times of Democritus and of Lucretius. The theory of atoms indicates that the molecules of a gas are undergoing some sort of a dance like the dusty motes one sees in a beam of light. In this dance, however, instead of being carried like the motes as passengers participating in a turbulence of some surrounding fluid, the molecules are subject to active forces emanating from one another according to a Newtonian theory of physics. The problem is to make this persistent dance of the molecules correspond to the observed facts of thermodynamics of the theory of heat-energy, and the other observed properties of gases. The theory of this dance is a statistical theory, in which the multiplicity of instances needed in every statistical theory (or, in other words, the multiplicity of the throws of the dice in a gambling game) is the actual multiplicity of the particles of a gas, or at least some closely related multiplicity such as that of the degrees of freedom in the motion.

Thus the multiplicity which is the basis of Maxwell's statistical theory is of the same sort as the multiplicity which occurs in the cosmology of Laplace. The next step taken in similar statistical theories of physics was made about 1900. Two names belong to this theory—that of the German, Boltzmann, and that of the American, Josiah Willard Gibbs, without any question the greatest star that has ever risen on the American scientific firmament.

These men, and more especially Gibbs, realized that physics must contain a statistical element far more deeply ingrained than that of the multiplicity of particles in a gas. Even such a system of a small number of degrees of freedom as the spinning top, as it follows out its orbit according to the Newtonian dynamics, goes successively through different positions in a space with enough dimensions to contain a complete characterization of its motion. If we examine this orbit at two different times, say time zero and time one, together with the momenta appertaining to them, the new positions correspond in a definite way to the old position-momentum complexes. In fact, we have a certain space of possible positions and momenta which goes on mapping itself upon itself with the progress of time.

There are certain quantities that remain unchanged throughout an orbit. In what is known as a conservative system, the most important of these is energy. The space of coordinates and momenta splits into different layers according to the choice of energy, as an onion splits into its consecutive shells. Each shell goes into itself with the progress of time, quite independently of what happens to the other shells. It was Gibbs's idea that the fundamental process of physics consisted in this continual mapping of one of these shells into itself. In this mapping he pointed out something that was already known to Lagrange, that there was a certain quantity on each shell analogous to area in two dimensions or to volume in three, such that this measure of a region was changed from instant to instant by the mapping engendered by the inner dynamics of the system. In this quantity, he saw something of the nature of probability, so that his probabilistic treatment of dynamics will be valid even for extremely simple dynamic systems as well as for those of the highest complexity, quite apart from any great multiplicity of component elements such as the molecules in a gas.

It was in introducing this intrinsic probability that Gibbs took the first step that has led to a completely probabilistic outlook in modern physics, and that even led beyond this to a simple probabilistic outlook in the modern theory of communication, of control engineering, of the theory of knowledge, and even in the social sciences.

This is one of these changes of intellectual climate which was bound to produce, and has produced, a flood of new discoveries and of new inventions in every conceivable field of human intellectual and constructive activity. I shall use the study of this change in climate to illustrate the relation of climatic change and climatic epoch to the history of invention.

For years, my interest in Lebesgue integration and in probability had prepared me for a wide extension of the Gibbsian point of view. In addition, I had been very near to the new quantum mechanism at its birth. This is a theory originated by Heisenberg in 1925, which was invented to correct certain deficiencies of classical Newtonian dynamics in the domain of radiation and of extremely small particles. I was much struck by the probabilistic elements which it contained, and by the doctrine of Heisenberg that certain quantities such as position and momentum were intrinsically incapable of an active simultaneous observation. These new ideas began to bear fruit when the exigencies of the Second World War thrust me into the problem of shooting ahead of a flying airplane and the consequent problem of predicting its course.

I sought to develop predicting machines of various sorts on a purely cut-and-try basis, and I found that what I was looking for would have demanded me to have my cake and eat it, too. I was studying the way in which the past observation of an airplane might lead to the computation of its future position. The accurate following of an airplane pursuing a smooth course demanded sharp and sensitive instruments; but these instruments, because of their very sharpness and sensitivity, were seen to be thrown out of action by every slight jar and by every corner of the course they were following. For very irregular airplane paths, the instruments I had suggested were inadequate, not in spite of their refinement, but because of their refinement. It occurred to me that this impossibility of achieving the ideal instrument all along the line had a close relation to the Heisenberg impossibility of observing at the same time where a thing is and how fast it is going.

The more I studied the problem, the more I realized that my difficulty was not a piece of casual malice on the part of a mathematical devil, but lay in the very nature of prediction itself. Thus I had to achieve a balance between refinement and ruggedness on the part of my instruments, and this balance I could only achieve by using my knowledge of the actual family of airplane flights for which the instrument was to be used. The theory of airplane prediction had become statistical.

Having this statistical theory already before me, I proceeded to do what many another scientist and inventor has done, and to ask what other problems I might solve on a similar theoretical basis.

I have before me a hectographed note emanating from *The Lamp* of September 1953. *The Lamp* is a house organ of the Standard Oil Company of New Jersey, and the paper is entitled "Serendipity (It's the Art of Being Lucky in the Laboratory)." The key word of this paper, which had been mentioned before by Dr. W. D. Connor, is the old-fashioned term serendipity, namely, the art of finding things you are not looking for. As the paper says, it was coined two hundred years ago by Horace Walpole, "the eighteenth century gentleman, writer and wit." Serendip is the corrupted form of the name Sinhaladvipa, or the island of Ceylon, as the Arabs wrote that Binhalen word.

There is an old tale of the three princes of Ceylon who, I quote, "as their highnesses traveled, were always making discoveries, by accidents and sagacity, of things they were not in quest of." This trait of the three princes, as the article shows, is a very vital weapon in the armory of the scientist. Science is fundamentally the art of getting into a closer rapport with nature, by questions and answers. In exercising this art, there is no reason whatever why the sole procedure should be to ask questions and to solve already formulated problems.

Indeed, the scientist who confines himself to this limited course of action is not using his brains at their highest efficiency.

After a certain question has been asked and answered, it will generally be found that the answer, if satisfactory at all, fits it not like a bathing suit, but like a loose robe; that is, a new method involves so much invention that it is scarcely likely that its value will be confined to the questions already asked. Under these circumstances, every scientist must occasionally turn around and ask not merely, "How can I solve this problem?" but, "Now that I have come to a result, what problems have I solved?"

This use of reverse questions is of tremendous value precisely at the deepest parts of science, but it is even important when it comes to particular engineering problems.

The art of serendipitous thinking led me to the study of the wave filter, the instrument to separate one message from another on a frequency basis. The theory of this was scarcely distinguishable from the problem of the predictor. Like it, it was based on a statistical theory. Thus I gradually came to realize the scope of statistics in my work and to apply them not merely to one communication-engineering problem, but to all. I was forced to see that the basis of all measurement of information was statistical, and that the frame for it had in fact already been provided by the work of Willard Gibbs.

Once I had alerted myself and the public in general to the statistical element in communication theory, confirmation began to flow in from all sides. At the Bell Telephone Laboratories there was, and is, a young mathematical physicist by the name of Claude Shannon. He had already applied mathematical logic to the design of switching systems, and throughout all his work he has shown a love for the discrete problems, the problems with a small number of variable quantities, which come up in switching theory. I am inclined to believe that from the very start, a large part of his ideas in communication theory and its statistical basis were independent of mine, but whether they were or not, each of us appreciated the significance of the work of the other.

The whole subject of communication began to assume a new statistical form, both for his sorts of problems and for mine. This is not the point for me to give the genealogy of every single piece of apparatus which this new statistical communication theory has fostered, but I can say that the impact of the work has gone from one end of communication theory to the other, until now there is scarcely a recent communication invention which has not been touched by statistical considerations. Thus, this whole widespreading branch of science represents a subtle working out of concepts implicit in Gibbs and in the Lebesgue-Borel team, but if I may say so, implicitly implicit, so that until some forty years had passed, no one could have seen the direction in which the earlier thought was bound to lead. This is, in my mind, a key example of a change in intellectual climate, and of the effect it has had both in discovery and in invention.

Circumstances Favorable and Unfavorable to Original Ideas

3

It will be fairly clear to the reader that the really fundamental and seminal idea is to a large extent a lucky and unpredictable accident. There was no absolute necessity for Euclid to develop the axiomatic theory of geometry, nor for Gibbs to insist so strongly on the notion of probability in thermodynamics. These innovations might easily have occurred somewhat earlier or considerably later, and are no more satisfactorily subject to betting about them, say, than about the particular house in the village which would next be struck by lightning.

For all that, though lightning is a sporadic phenomenon even for good betting, we do have a general idea of what circumstances are favorable for lightning and what are unfavorable. We do not build a house on top of a high and isolated hill without being particularly attentive to our lightning rods. So too in matters of invention, occasional and sporadic as the phenomenon is, we may look to certain circumstances to favor it, just as we may look to other circumstances to cut down the risk of lightning.

There are certain procedures which are undoubtedly favorable for invention and discovery. One of the most potent tools in reanimating a science is mathematics. To some extent, a mathematical treatment of a science consists in writing down its data and its questions in a numerical or a quantitative form, but it is perhaps better to consider that here number and quantity are secondary to a logically precise language. If a certain question is to be asked in biology, and if we are to ask it in biological language, then we ourselves and whoever reads our work are likely to be strongly conditioned to think of what we have done as the answer to a biological question. However, if we express our ideas in a mathematical form, we are using what is much more likely to be a colorless and indifferent language. Just because of that, we are far more likely to recognize the same question even if it is asked in a totally different field. This greater scope is far from of trivial significance.

Let me give an example of the cross-fertilization of disciplines by the mathematical expansion of the ideas contained in one of them, and their embodiment in a neutral form extending across disciplines. Suppose that we are studying the course of an epidemic of measles. Measles is a childhood disease, and to say that a childhood disease is one of children very often amounts to statements concerning its nature and behavior in which the word "child" does not occur. In the first place, measles is highly infectious to those who have not the natural or the acquired immunity. This means that a large part of the population, if exposed to measles, would get it on the first occasion of exposure, which is likely to be quite early. An attack of measles itself conveys an immunity which is high, if not perfect. Therefore, in a population which has largely been exposed to measles, the older members are probably immune either by birth or by new infection.

Now, measles breaks out in sharp attacks with an approximately regular distribution over time. Can we see anything in the nature of the disease that tends to favor this? The answer is that an epidemic means a transfer of the disease through a population in which immunes and nonimmunes are intermixed, by a chain of infectious contact running between nonimmunes. The length of these chains is intimately connected with the probability of an epidemic and with the extent to which it will rage.

It is a mathematical fact that when we have any sort of chain of contact between individuals, the length of the chain of contact between a particular set of individuals such as nonimmunes will depend in a very critical way on the percentage of these particular individuals in the community. Thus, if there are only one percent of nonimmunes, then on the average only one contact in one hundred of a nonimmune is likely to be with another nonimmune, and if we go still further, we shall see that the expected chains become short and incapable of spreading. On the other hand, if seventy-five percent of the community are nonimmunes, the chance is very high that nonimmune chains will run from one end of the community to the other, and the stage is set for an explosive epidemic.

So far we have been discussing a medical situation in a quasi-mathematical language. Let us see if this language is equally applicable to widely different situations. If we have a mixture containing hydrogen and oxygen in explosive proportions, and also a neutral gas such as nitrogen, the question of whether a flame can sweep through it will depend on the problem of whether one hydrogen molecule will be sufficiently near another so that its combustion will excite the combustion of the other, and of the length of chains of hydrogen molecules with this type of contact. If the chains are long, then we shall get a mass explosion, which is the mathematical equivalent of an epidemic. If the chains are short, a local reaction will not spread.

Notice the following: If we have a hydrogen and oxygen mixture heavily diluted with nitrogen, and try to start an explosion in it, nothing will happen. If we let the oxygen run in separately in proper proportion and displace the nitrogen, at some stage the chain length will have built up until we get a puff of explosion, if there is any pilot. This puff of explosion will reduce the percentage of hydrogen and oxygen so that no further explosion can take place until enough new hydrogen and oxygen have arrived to significantly increase the length of the chain.

Now compare this with the measles epidemic. Where a measles epidemic has taken place, practically all the nonimmunes who have been exposed to measles will catch them and will then either die or become immune. Then for a considerable period no new epidemic can start. As, however, the proportion of nonimmunes in the community increases with the birth of new children or with the moving in of strangers, a level will be reached where the community is ready for a new epidemic, and it is safely predictable that within a short time some spurt of infection will set the chain going again. Thus in the cases of both the epidemic problem and the explosion problem, we shall have a series of puff outbreaks at fairly definite intervals separated by periods in which an outbreak is impossible or unlikely.

The advantage of a mathematical description of this phenomenon is that our attention is not specifically turned either to measles or to combustion, but remains in an indifferent state equally applicable to both. But there are even wider problems whose solution is favored by putting this same complex of questions in an abstract form. For example, let us have a glass tube with electrodes at each end, and let the tube be filled with a mixture of steel ball bearings and glass beads. Will the tube behave more as a conductor or more as an insulator? The answer depends on something very similar to what we have already discussed—on the question whether a chain of steel balls in contact is long enough to reach from one electrode to the other, or, on the contrary, is broken up into subchains by completely separate regions of glass beads. The same mathematical concepts which help us to state and to solve the epidemic problem and the flash problem are equally relevant here. In general, the same concepts are available for discussing the conductivity and other properties of an aggregate which, though mixed, may not be macroscopically mixed. An alloy between one or more metals will form such an aggregate.

Mathematics allows us to state the essential and to bury the issues of the inessential. It allows us to ask the same questions in many fields without committing ourselves to one of the fields in particular. In the hands of a mathematician who is sensitive to the possible interpretations of his mathematical language in many different fields, it is a powerful organ of invention and discovery.

Yet a mathematician may fall short of the best use of this talisman in at least two ways: On the one hand, through inexperience or snobbery he may be so much of a pure mathematician that he fails to be able to carry through that step of interpretation in concrete terms which is an essential counterpart of his purely abstract thinking. On the other hand, we have the opposite weakness of the applied mathematician who has become a master of the art of universal mathematical translation but who, once this is accomplished, has an inadequate and out-of-date repertory of purely mathematical tools. The best use of mathematics as an aid for discovery can be made by the man who does not commit himself to either of the two labels, pure and applied, but who is willing to combine the mathematical resources of the pure mathematician with the translating ability of the applied mathematician.

Many times we, in solving a scientific problem by mathematical methods, shall run into the type of abstract difficulty to which the mathematician refers either as a difficulty of rigor or as a difficulty in his existence theorems. These difficulties have generally been considered as the appropriate field of the pure mathematician. The assessment of the relation of such mathematical difficulties of a problem requires the union of the highest abilities of the pure mathematician and those of the applied mathematician.

On the one hand, a careful study of the problem may show that these difficulties belong not to the problem itself but to a specific way of approaching it, and can be sidetracked by a practitioner of wider vision. On the other hand, a mathematical difficulty may be a true signpost for a physical difficulty which cannot be turned aside, or which can only be turned aside by a very radical modification of the physical conception. An example is readily at hand.

In the statistical study of waves on the surface of water, as for example in the case of a broken sea, certain failures of convergence will be observed which may seem purely mathematical on superficial examination, but which have the following meaning: If the sea is even moderately high, the water wave will not continue to be a surface met by a vertical plumb line at only one point, but will curve over. One part of the wave is below another, and the wave will eventually break. To make a statistical theory which is not cognizant of this breaking can scarcely be fully adequate in any case, and must be grossly wrong in the case of a high sea. Yet without a high degree of sophistication as to what is really taking place on the abstract level, the danger signals implicit in the mathematical failure of such a theory may escape our reading.

So far in this chapter, I have been discussing those aspects of the intellectual environment that belong in a sense more to science than to the scientist. There are, however, certain matters of the intellectual environment that belong to the scientist as an individual.

Since the individual scientist is very deeply conditioned by the social milieu in which he lives, these aspects of the creative environment are not easy to separate in any sharp way from corresponding aspects of the social environment. In placing them in this book, I shall have, to a certain extent, to treat them in both places, for the sort of individual we get has an internal relation to the sort of individual whom society demands. There will therefore be a considerable amount of repetition between chapter and chapter, but I shall make my basis for the separation of these two points of a closely related subject the question of whether I am primarily considering the individual type of the scientist as such, or the availability of this type at present and in past ages, as conditioned by the environment.

Here I run into the difficulty that a well-organized book may be a pedantically organized book, and that in no way can I discuss either of these subjects without bringing in much material that belongs to the other chapter. I can only hope that the reader will bear with me in this matter, and will be aware that I do not have before me any ideal choice.

A source of the spirit of invention or discovery that has been peculiarly potent over the last four centuries is the modernization of the old spirit of the linguistic, historical, and literary scholar. When the Renaissance came to Europe, there was an enormous sense of new doors being opened and old doors being reopened. The documents of the ancient world which, with their owners, took refuge in Italy at the fall of Byzantium represent a mass of material then for the first time accessible to a civilization undergoing its growing pains, and they were a challenge to every man of intellectual power.

The repercussions of this factual material on a Europe already not unfamiliar with philosophical speculation created the new class of the humanists. At the beginning, the scholarship of the humanists was primarily literary and classical, but the existence of a disciplined element in society used to the effort of prolonged thinking and study furnished the medium in which the new science could be created.

The humanists were not the only European repository of learning, nor the only one to be thrown open to creative scientific thought. The Jewish Talmudists represented a parallel stream. This stream, however, had largely been cut off from that of the humanists by mutual prejudice and difference of customs. It is true that many humanists were in contact with Talmudic scholars, but this contact was perhaps sporadic and exceptional. When, under the influence of Moses Mendelssohn, the sporadic contact between the Talmudist and the Western scholar was changed into a vital channel of Jewish learning, the intellectual power and the intellectual discipline of the Jew was made available not only for literary learning but also for modern science. It does not need a Jewish chauvinism to point out that since that time the army of scientific learning has been reinforced by a valuable and even indispensable new contingent.

The Jews represent an accrual to Western science of what we may call a piece of the Orient on Western soil. The last two generations have seen a duplication of the process by which the learning and intellectual discipline of the East have come to the aid of the West within the truly Eastern countries themselves. The Confucian scholar-statesman and the Indian pundit have alike transferred their drive to scholarship and their training in scholarship to the newest learning of science. Again and again, it has become clear that the qualities inculcated in an ancient scriptural learning need but little modification to serve the curiosity of the latter-day physicist or biological investigator.

However, while the East is coming to the reinforcement of the science of the West, the West has suffered, and is continuing to suffer, a drying up of the springs of learning which have rendered modern science possible. Two wars have all but removed from the lists the Germany which was the chosen home of nineteenth-century learning. Neither the postwar austerity of England nor the confusion of France has been particularly favorable to the continuation of their ancient traditions of scholarship. The two new countries which cast the longest shadow in the science of the present day are the United States and Russia.

In each of these two countries, the European tradition of scholarship has had to survive in a climate in which scholarship, for all the favors it receives, has been very greatly subordinated to other interests. In neither country does the man in the street care very much for the classical learning out of which all Western European scholarship emerged. In each, the scholar is primarily valued as the servant of interests which are not in themselves scholarly. I do not wish to belittle the great expenditures of money and of effort which have been made both in Russia and in the United States to secure a continual supply of scholars to the commonwealth, but in neither country has the work of the scholar been the main consideration. In Russia, the scholar may be left reasonably alone in his scientific work if he has amply proved that he is not a danger to the community, but it is this state-community and the furthering of its social and economic principles which must in the end override all claims of the scholar as such. They have reduced the notion of scholarship for scholarship's sake to the same heretical level to which the concept of art for art's sake has already been reduced.

In America, the scholar need not yet prostrate himself before a political theory, although this reduction of his stature may not be far off. Yet he is expected to assent to the deification of the businessman and of the order for which he stands.

Today as I look through the papers, I see that the tax-free national foundations, such as the Rockefeller, Carnegie, and Ford, are under attack as hotbeds, if not of Communism, then of Communistic ideas. In the jargon of the present day this means simply that they are not inflexibly devoted to political and scientific orthodoxy. We have seen in Russia that the atmosphere of demands for political orthodoxy must hamper even a rapidly developing scientific culture. What can happen in Russia can happen here, and the general attempt to channel the wayward streams of thought by rule of thumb may well be expected to end in lowering the intellectual water table and turning vast areas of the soul which need our cultivation into dead and useless deserts.

Let me give an example of the layman's contempt for the scholar. Many years ago, when my father was traveling in Russia, he happened to be in the same train compartment as a *schechter*, a Jewish ritual butcher. The *schechter* was expatiating on the merits of his position and on the good profit that he made. At length he asked my father what his trade might be. My father explained that he was a professor at a university. The *schechter*'s answer was, "Auch ein gutes Geschäft!"—"That's a good business too!" I am afraid that either in modern Russia or in the America of this generation, this represents about the highest evaluation which we scholars are likely to find at the hands of the community.

The point of this story is the low evaluation that the *schechter* was putting on purely intellectual work. The evaluation which is put on intellectual work in our American community is scarcely higher. For years in Europe and in the Europeanized America, the businessman and the shopkeeper smarted under such snobbish epithets as "counter-jumper," "boutiquier," and "epicier." He has equally smarted under the pressure not only of Communism but of all the different shades of socialist movements.

Within the last decade or two, he feels that he has come into his own, and that he holds in his hands all the aces in the game of the development of America. At one time, he was content to urge capitalism as a permissive theory, as the assertion of a claim that he had his due and dignified place in society, and that he must not be passed by either by the snobbery of the aristocrat or by the snobbery of the intellectual. At present, there are signs that he is turning his demands for room to live into demands that his way of life be recognized as the basis of life of the entire community and that those who are not in accord with him should be punished with the whips and scorpions of a new Inquisition.

This new orthodoxy does not constitute a climate favorable to devotion and consecration. Devotion in business there is and has been, and it was probably stronger under the old shopkeeper business dispensation, which may have been tolerant towards the businessman as to a rather shabby treatment of his customer, but was adamant when it came to the question of his fulfilling his business obligations. Many an old-fashioned businessman has committed suicide rather than face bankruptcy, not so much because of the enormous effort of going back to the beginning and starting all over again, but primarily because, as he conceived it, there was a mark on his good name that nothing could erase.

I will not say that this type of integrity has vanished in business, but merely that in the balance between integrity and aggressiveness, the needle has shifted over pretty far in the direction of aggressiveness. The devotion inculcated by the church is seriously involved in this, and for many a successful man, the church has turned itself into a black police for the protection of the existing social order in all its details, rather than an institution where he is called upon to repent and to do penance. Even though a certain conventional respect for the clergyman and his need for devotion has been left within the present social order, outside of this limited range of permission, there are many who have come to regard devotion as not merely ridiculous, but as potentially dangerous.

The businessman is brought up sharp by meeting the scientist who cares more about his science than about the tangible rewards of science. He feels that he has come into contact with a smooth, hard, impregnable fortress of the soul, in which he can find no obvious point of attack. To him the scientist or the man of letters represents a core of possible defiance. In this threat of defiance he sees something to be eradicated and stamped out. The less that the free-lance individual demands of him in property and the riches of the world, the less sure is the businessman of his own final dominance.

Thus the powers that be are very glad to see that in the new generation there is a trend away from self-sacrifice, from the passion for learning, and from all these uncontrollables. They are delighted that the doctors are showing an increased interest in the arts of spreading their practice and collecting their bills. They derive a very real satisfaction from the fact that the young engineers and scientists are flocking to their laboratories and leaving the universities and that the institutions of pure research are understaffed and dangerously weak. Those scientists who cannot talk in sums of money less than a million dollars are his own men, and he encourages them, perhaps not to join his own country clubs, but to join country clubs of a slightly smaller prestige, and to buy, not Cadillacs, but the precise make of car that would show at the same time a proper deference for his own superiority and a proper worship of his own ideals.

What he fails to see is that he is paying for the immediate subserviency of the scientist in the inability of the scientist to furnish the long-time and deep-lying developments on which the community as a whole and he himself in particular ultimately depend. For the megabuck scientist, a really deep study of the laws of nature demands at least a temporary retirement from the ranks of scientific management, and many hours of contemplation in the peace of his own study, before anything emerges which will bring him spectacular attention.

This is a gamble which he cannot afford to take. It withdraws him from the very intense struggle for top position in which he has been indoctrinated. If for one moment he fails to watch what is fashionable and what the other men are doing, he invites another climber to push him off the ladder. Thus the world of competition and of self-advancement can only continue to exist by virtue of the existence somewhere within it of a corner in which scientists are not self-seeking, and only compete in a deep joint effort to disclose the secrets of nature.

If we want to discover or develop the true scientist, we must begin early in the childhood of the man who is to become a scientist. The man who is to be devoted must have had a chance to see what devotion is. The man who is to be governed by a desperate curiosity about nature and by an unwillingness to be baffled by anything that he can possibly overcome, must form this orientation early, before he is caught by the better-paying worldly values. Our schools must teach something more than conformity, and must demand something more than well-rounded nonentities. Whether the courses of the high school be in modern languages or classics or mathematics, they must recover a part of the bite and weight they once had. Short of this, our civilization will drift into a Byzantine mediocrity, and our science will be governed by officials or employees, not by men.

The Technical Climate and Invention

4

Invention, as contrasted with the more general process of discovery, is not complete until it reaches the craftsman. Beside the conditioning of new inventions by the ideas on which they depend, there is a further conditioning in terms of the materials and the processes available. The limitation of the effectiveness of invention by the present availability of materials and techniques appears in a very conspicuous form in the study of the notebooks of Leonardo da Vinci.

What cannot fail to impress anyone who studies Leonardo is the relatively large use which his techniques make of wood and leather. Both of these are materials for which an adequate technology existed in his own time, whereas the technique of metals was decidedly limited. The arts of the founder and the blacksmith were well-known and highly esteemed, but they were by no means adequate for the production of precise mechanical forms. The surface of a cast metallic object is not very precise, as certain of the metals tend to expand in cooling and others to contract. The first tendency requires the use of sand molds, which cannot be given too sharp a surface, whereas the second tendency pulls the metal away from the mold. Moreover, cast metals generally have a crystalline structure with the grain not too relevant to the stresses to which the metal part will be finally subjected. They often are brittle. Under the more or less primitive conditions of the Renaissance, they were far from adequate for the precise work of the engineer, as distinguished from the aesthetically precise work of the statue-founder.

The hammer of the smith is an even less precise tool than the mold of the founder. It does, indeed, have the advantage of giving a piece of metal with a known grain, and can be supplemented by the process of tempering and the related process of annealing. However, before the work of the smith can yield us an accurate finished surface, it is usually necessary to call in the kindred yet different art of the locksmith. This art of the locksmith was well known long before the time of Leonardo, but did not reach its full flowering until much later.

The original tool of the locksmith is the file. With the aid of that instrument and of the kindred saw, the locksmith is able to cut out the complicated shapes of key and keyhole and tumbler and bolt, and to produce surfaces with a sufficient precision to clear one another and yet to engage one another in the action of a lock. The craft of the locksmith has continued to the present day, and in modern Germany, or at least in the Germany of a generation ago, this craft was considered to embrace the crafts of the machinist and the toolmaker. The German professional machine tool operator had been apprenticed as a locksmith, had served as a journeyman locksmith, and, if he aspired so high, had been graduated as a master locksmith, by producing an approved, finished piece of work: in the technical sense, a masterpiece.

Why should the machine tool operator trace his spiritual ancestry to the locksmith rather than to the bellfounder or the blacksmith? It is because the locksmith's art is that ancient craft in which metal is shaped by removing parts of it. What the file did in the old days, the arts of the cutting tool on the lathe or the planer or of the lapping of metal by abrasives have done in later ages. Yet these modern cutting or grinding tools had to wait until richer sources of power were available than those which could be found in Leonardo's time, and the work of the Renaissance and medieval locksmith was notoriously arduous, slow, and exacting.

It is no wonder that the craftsman of the time, including Leonardo himself, turned to the kinder, softer materials of wood and leather. Leather is a useful and flexible, airtight and watertight material, but is in no sense a suitable material for precise craftsmanship. Wood and its kindred materials, ivory and bone, are no contemptible mechanical materials. They can take a high polish and in some cases furnish a sufficiently resistant surface to replace metal. However, the woods which are suitable for such an employment are in general tropical materials such as teak, ebony, or olive wood, and themselves require the file of a locksmith to bring them to precise shape. In Leonardo's time, these harder woods and woodlike materials were no more amenable to manipulation than were the metals.

The more ordinary woods, rather soft or moderately hard, could indeed be worked to a precise shape by the saw, the chisel, and the auger, but the price for this greater workability was softness and often weakness. A material which works easily abrades easily, and cannot be maintained to accurate shape without frequent replacement. Moreover, wood is not a substance with which it is easy to minimize friction and the waste which friction incurs. To see one of Leonardo's drawings and to imagine it in action is to see a flimsy, creaky machine which will indeed turn over, but not with any willingness, and whose working surfaces will soon be polished by slippage and dented to the point where they are ineffective.

I do not know at what period it was discovered that friction between metallic surfaces could be minimized by the use of oil and grease, but it probably was found out relatively early, and most likely through the problem of the design of the cart axle. We now know that wood, too, can be lubricated, although it is not well suited for oil and grease. Soap and water are among the best lubricants for wood, but I cannot say when their suitability was discovered. In any case, while many of the devices of Leonardo could work and could be made to give a moderately competent performance with proper lubrication by soap and water, their feeble performance would break the heart of any modern engineer.

The precedents for Leonardo's machines were of course to be found in the watermill and the windmill. The watermill, which in Roman days had nothing better to use than the inefficient undershot wheel, had developed the overshot wheel at the end of the Dark Ages, and this wheel is quite reasonably efficient, although not suited to the exploitation of very considerable quantities of water. The windmill was even less suited for the obtaining of anything but driblets of energy. Something more than mill techniques was needed before Leonardo would come to his own.

The path which ultimately led to the everyday realization of the mechanical gadgets of Leonardo led away from wood to the properly finished metal, first of the locksmith and later of his spiritual descendants, the watchmaker and the scientific instrument maker. The medieval maker of tower clocks was probably not much more than a glorified locksmith himself. The maker of "Nuremberg eggs" was indeed a consummate craftsman within his limitations. He borrowed the notion of the mainspring from the steel technique of the swordsmith, and thus took a first important step in the storage of energy. It was with the seventeenth and eighteenth centuries, with the rise of the pendulum and the hairspring, and with the smaller and more precise devices demanded by the navigator for his chronometers, that the art of the clockmaker and the watchmaker came to its own. The lathes and elementary machine tools used in the construction of these smaller instruments were sufficiently small in size and demanded sufficiently little power as to be run by the hand or the foot of the artisan. These lathes were the genuine ancestors of the machine tools of the present time, not only in their purpose and in their conception, but also because, as we shall see later, they served to finish bit by bit the gradually increasing sizes of metal parts needed to fulfill the demands of the Industrial Revolution.

The seventeenth century of science was fortunate in being the great age of optics. The craft of forming optical glass is almost an ideal one for the man of scientific inclinations, and this is quite apart from the usefulness to the scientist of the products of this art. There is no other field where so much precision can be reached with so few tools.

Of course, the art of lens grinding is older than the seventeenth century. The Arab, Alhazen, greatly advanced the science of optics, on which it is based, but we have no evidence that he used ground lenses, and the art of the lens grinder began with the spectacle makers of the later Middle Ages and the Renaissance. Glass itself is far older still, and indeed may have been an accidental discovery of primitive man when he built his fires on the seashore or in the desert, when his ashes and the sand may have fused together into translucent lumps. Glass is thus very old as a vehicle of art, but it had a long way to go before it came to be a proper material of a scientific technique. The early forms of glass were flawed, colored, and semi-opaque, and although glass became a material that could compete with some of the uses of pottery at a fairly early epoch, it took many centuries and even millennia before it could be made clear, reasonably colorless, and in pieces of adequate size for further work. Moreover, of the many techniques by which glass can be handled, the early ones involved melting and blowing and welding, and were quite unlike the technique of grinding which made the construction of precise optical instruments possible.

This technique of grinding by abrasives and of polishing by rouge and other fine and mild abrasives is, of course, something of a counterpart to the metalworking craft of the locksmith, of which we have already spoken, and became joined with it in the closely related art of the construction of accurate mirrors of metal. This, indeed, was but a development of another very ancient craft known both in the East and in Greece.

To grind a plane mirror, as all amateur telescope makers know, requires no tools except three pieces of glass, the pitch to hold them, and the proper abrasives and polishing materials. If we grind plate A against plate B, and they continue to fit after their respective orientation has been changed through small stages, it is mathematically clear that they must be either plane or spherical. If we take plate A and grind it against plate C, it is unlikely that they will fit on first contact, but if they do not, further grinding will tend to take the high places off each, and thus is likely to leave plate A flatter than it was before. Thus, by a continual sequence of grinding A against B, B against C, and C against A, we can expect that on the whole we are removing high places and flattening the plates. If we finally reach such a stage that A fits B, B fits C, and C fits A, the only possibility is that all three will have been reduced to planes. Thus, almost without tools, we shall have produced a flat standard which will meet more than the sharpest requirements of the modern toolmaker.

The same methods which produce flats produce spheres, if we only grind two plates together, and if we guide the end of our stroke by overshooting the edges in the proper manner. Having obtained our spheres, we still may wish to modify our surface so as to have better geometric properties for the concentration or spreading of light. In this task, the optical properties of light are themselves the best tool. We test our surface by refraction and reflection. This is why the making of telescopes remains a frequent pastime of the man who combines intellectual interests and a fair amount of mechanical skill but who cannot afford an elaborate kit of tools.

The earliest telescopes were lens telescopes, and the microscope has almost always been a lens microscope. The general principles of the grinding of lenses and mirrors were the same. Lenses, however, are more difficult to grind than mirrors. They have two surfaces instead of one, their transparency is imperfect (which puts an upper limit to their effective size), and they involve a homogeneous, almost flawless material. The telescope of Galileo was a lens telescope, while Huyghens used a mirror as his principal optical element.

One advantage which the telescope maker had was the relatively fine metal workmanship of the maker of astrolabes and other lensless astronomical instruments before him. This same metalworking craft was needed in the telescope and microscope to supplement the work of the lens maker and mirror workers. A telescope is, after all, not merely a structure of lenses or mirrors, but also must have a casing, which is generally of metal, and the screws and racks and gears which are necessary for its adjustment. Thus the profession of the lens grinder very soon spread into the profession of the scientific instrument maker, which, of course, involved the locksmith-like art of the metalworker at its highest levels. Moreover, the new discoveries made in mechanics by Galileo and Huyghens had set the stage for a complete reformation of the art of the clockmaker, and had replaced the crude but ingenious work of the maker of tower clocks by a new technique of pendulums and mainsprings which combined the theoretical work of Galileo and Hooke with the previous cut-and-try methods. The art of the optical instrument maker and of the clockmaker were, in fact, closely connected from the beginning. In the eighteenth century, these two convergent crafts were to contribute almost equally to a new revolution in navigation.

It was entirely natural for the development of optical instruments, the mathematics of optical instruments, and the developments in the technique of constructing optical instruments to go hand in hand. When this progress was made at just the time in history when optical instruments were most needed, it is not surprising that a new age of invention was inaugurated. Galileo's telescope appeared just when the cruder instruments of Brahe and Kepler had exhausted their possibilities, and Leeuwenhoek's microscope could not have come any later without a serious handicap to the continued progress of physiology and anatomy into the realm of the small.

Thus the seventeenth century represents a most surprising marriage of new thought and new technique. It was a time in which a philosopher like Spinoza could reinforce his independence of thought with the economic independence of a lens grinder. Either one of these would have marked it as a great age of discovery and invention, but the two together are quite sufficient to constitute the beginning of a new period in which technical progress ceases to be sporadic and is thoroughly integrated into our civilization.

The sailing ship is now so dead that we are almost beginning to forget what a magnificent craft it was and how much it demanded in the way of its own sort of engineering. It is true that the structural improvements of the ship and its gear came on gradually, that it took centuries for the steering oar to give way to the tiller and rudder, and for the tiller, which was at first manipulated directly by hand, to be controlled by the whipstaff or (at the end of the seventeenth century) by the wheel. The square foresails and the lateen mizzensails of the Elizabethan ships made way, in the course of time, for the jib and the flying jib on the one hand, and for the square-rigged mizzen on the other. But these improvements in the vessel and in the material of seamanship were much slower in making their appearance than the corresponding developments in navigation itself.

Before the beginning of the eighteenth century, latitude indeed was easy to determine, but longitude almost impossible to determine. A ship would sail along a coast until it had made enough northing or southing to come to the parallel of the desired landfall on the other side. From there it would sail due East or due West until it made its landfall. If the captain were so unlucky as to fail to make a correct estimate of his winds and currents and consequently of his dead reckoning, he would reach the opposite coast earlier or later than he had expected, often with disastrous consequences to his vessel. Thus both England and France, the great countries of the sea, found it necessary, in the early eighteenth century, to offer prizes for adequate methods of determining latitude. These prizes were repeatedly won in both countries, and the rewards awarded by each to the men who won the prizes were shabby in the extreme.

The problem of longitude is one of clocking. Its theory had long been understood. Two practical methods were known to be available. One was the construction of mainspring chronometers sufficiently sturdy and self-compensating to endure the motion of the ship for days without being thrown out of action. The other was the better observation of that natural clock, the moon. Both methods were ultimately successful, and both involved great improvements in instrumentation. The clock had to be constructed to an accuracy and a sturdiness not yet contemplated, while such unwieldy apparatus as the telescope had to be evolved into the convenient and portable mariner's sextant. In an age of navigation, the great industrial techniques came to be those of the clockmaker and the optical instrument maker, and in both of these, new levels were reached in the manipulation of metal and of glass.

By the end of the eighteenth century, and in some matters much earlier, the traditions and the tools of the clockmaker and the scientific instrument maker reached levels comparable with those of modern engineering technique. The lathe had become adapted for precise use on brass, and proper dividing engines had been devised for calibrating and marking line and angular scales. In no other fields of work except perhaps that of the borer of gun barrels was the machine tool even approximately so well developed. Thus it is no accident that Watt and many of the other inventors of the time were originally trained as clockmakers or scientific instrument makers.

The steam engine as it originated was sadly imperfect. It was an instrument for low pressures. In fact, in the form devised by Newcomen for the pumping of mines, it worked between a vacuum and a pressure scarcely, if at all, greater than atmospheric.

The Newcomen engine did not involve engineering of any very high level. The engine of Watt went beyond the Newcomen engine in several important respects. For one thing, the valve gear of the Newcomen engine which regulated the admission of steam and the exhaust was crude in the extreme. At the beginning, it had merely consisted in a cord to be pulled by an attendant. Later on, the ingenuity or, as it is said, the laziness of one of these attendants resulted in a crude device by which the motion of the piston itself pulled the cord. Watt's engine, on the other hand, possessed a fully developed valve gear of a considerable degree of complication.

Thermodynamically, the Newcomen engine was grossly inefficient, inasmuch as it used the same vessel as cylinder and condenser, thus wasting an overwhelming part of the heat energy. Watt invented a separate condenser, but what was equally important, he had designed his engine with the purpose of turning rotating machinery, rather than of merely operating a pump rod.

Thus, while the Newcomen engine made no excessive demands on the technology of the eighteenth century, Watt's engine strained this technology to its limit. For example, for efficiency it was necessary for Watt to have a well-fitting piston, but the best standard of workmanship to which he could appeal permitted nothing more than a piston that fitted its cylinder to the extent that a thin sixpence could just be squeezed into it all around. Watt made an extensive use of gear trains and linkages, at a time when gear teeth had to be cut by hand, one by one, and when his linkages could not be expected to be more accurate than good blacksmith's work. In fact, Watt suffered under the need of developing a technique for working metal, parallel to the technique of the clockmaker, but capable of taking work of ten times the dimensions. This technique had to be developed from the bottom up.

The paradox which faced the engineers of Watt's time was that the precise mechanism of the steam engine could only be constructed with any ease by tools sufficiently powerful to need a steam engine (or possibly a water mill) to drive them. In all engineering, there is a certain family history, a certain genealogy. The smith's hammers were forged by the hammers of an earlier smith. Thus, when the Industrial Revolution began, the first difficulties of construction were faced boldly and brutally by hand methods. Each new step then made the taking of later steps easier and more accurate. The hand tool made the first steam engine, and this steam engine made the tools for the construction of later steam engines. In the course of this growth there was a successive increase in the size and the strength of the parts which could be made, and the machine tools of the present day thus have grown up by the growth of generations from the small lathe on a clockmaker's bench.

One country which has had a peculiarly important part in the history of early invention is China. To China we owe the discoveries of printing, of paper, and of gunpowder, besides great developments in the techniques of textiles and of metals and those methods of well-boring which are the direct ancestors of our art of drilling oil wells.

Before I go into these inventions one by one, let me say that the Chinese social ideology was very favorable to craftsmanship. The position of the craftsman in China was about as high as it later became in the Flemish and Italian city-states of late medieval Europe and the Renaissance. In Europe the soldier has always stood well towards the top of the social scale. In China, the Confucian classification gives the highest rank to the scholar-statesman, the Mandarin. After him follows the farmer, including all the levels from that of the ploughman to that of the country gentleman working his estate. Next comes the artisan, and below him the merchant who sells the work of others. Still lower come those who perform the various menial tasks of society, and at the very bottom the soldier. "The soldier is below," says a familiar Chinese proverb, and another one states that "you don't use good iron to make nails, and you don't use good men to make soldiers." Thus the craftsman stands perhaps a little above the middle rank of Chinese society.

The conservatism of nineteenth- and early twentieth-century China is not a phenomenon of long standing. It is probably about contemporary with the overpopulation of China, which did not date much further back than the end of the seventeenth century. Up to the time of the Renaissance China was clearly superior to Europe in intellectual activity, in the number of people kept alive, and in the amenities of life. This is made obvious to us by what Marco Polo has to say about the country. Indeed, even a man of the European Renaissance like the Jesuit Matteo Ricci, who came to make converts and found many things in China repugnant to him, still had to admit that a considerable number of details of Chinese life in the sixteenth and seventeenth centuries were more civilized than what he had found at home in Europe.

Indeed, Ricci's own career shows that the China of his time was far from an intellectual vacuum, and that the problems of the intellect had a great fascination for the Chinese rulers. Ultimately he became a Mandarin like Marco Polo himself, and a fairly good Chinese scholar. What had endeared him to the Emperor and the high Mandarins was the knowledge of Western mathematics and astronomy which he brought to China, and which from that time on became a part of Chinese higher education. As one example of his scientific proficiency, he mapped the world for the Chinese, choosing a projection which was legitimate but which sufficiently exaggerated the relative area of China to be acceptable to Chinese prejudices.

Let us take some of the individual inventions by which China is known in the West. As to the inventions of printing and of paper, we generally consider these in the wrong order, attributing too much importance to printing and too little to paper. The problem of good writing materials has troubled man from the beginning of writing itself. Early writing consists in inscriptions on stone or pottery or metals, and this was the case in China itself. The countries which have led in civilization have almost always found light and portable writing materials on which they could keep their records. In Babylonia and the neighboring countries, the material was simply baked Euphrates mud. A system of imprinting this with a wedge-shaped stylus led to a culture with many records and on a level extraordinary in its time. Early India and the countries further to the East made much use of the palm leaf. This, by the way, suffers from an excessive grain in one direction so that it splits easily and demands a form of writing with more or less circular strokes. We find this circular writing in Ceylon and southern India down to the present day. The papyrus of the Egyptians and later of the Greeks and Romans was made by pasting together successive layers of the pith of the papyrus reed. This substance has a less marked grain, but it is still a fragile and brittle material.

Parchment and vellum were tougher and more permanent writing materials. They suffered under the disadvantage of excessive cost, so that to the expansiveness of the material there became very properly joined the additional expense of fine, slow, and artistic methods of writing. These were quite suitable for religious scrolls and for an accomplishment which belonged almost exclusively to the priests and to the rich, but made the general use of reading and writing most difficult.

It was China which discovered the material which is almost universal at the present time, which is grainless, cheap, and abundantly available, and which we call paper.

The Chinese even to the present day have a peculiar respect for writing. They do not like to throw away any fragment of paper with writing or printing on it, and they even make mural decorations out of copies of writing and printing on scraps of paper. It is natural to suppose that they must have had an enormous emotional pressure from the beginning to find a material which would make writing easy.

Apparently the first paper that they made was not a true paper in the sense of being constructed of cellulose. It was a felt of silk fibers analogous to paper in its structure but not in its chemistry. Later on, the Chinese used mulberry bark. This gave a true cellulose paper, and was followed by a paper of cotton fiber. The resulting material was cheap as compared with parchment, and perhaps even almost as cheap as papyrus. It was not as tough as parchment, but tougher than papyrus or palm leaf.

The Chinese mode of writing with brush instead of pen made rather lighter demands on the toughness of the writing material than an instrument with small writing point and more likely to cut into the writing material. Associated with this brush mode of writing, which soon became a brush mode of painting as well, was the use of that rather viscous ink which we misname "India ink." The paper material became very soon quite as much a material for painting as for writing. This is not remarkable, as the Chinese themselves scarcely distinguish the art of painting from the art of calligraphy.

With an abundant writing material at hand, and with a viscous ink, printing originated almost automatically. All countries have used seals as a way of identifying documents, but these seals have been of various sorts. Babylonian seals either were stamped in clay or were inscribed cylinders that were rolled in clay. The seals of the West have often been in wax, although certain very special seals, such as the bulls of the popes, were impressed in thin sheets of lead. From the beginning, however, the Chinese have made great use of ink seals, and here the viscous character of their ink comes to the fore again. These seals have often been small ivory or stone objects carried by the owner for the personal identifying of his documents. Indeed, the writing of Chinese is so preordained and organized that a signature by brush is scarcely enough to identify its maker. Chinese banks, until recently at least, have demanded that checks made out in Chinese brush writing be attested not by the signature of the maker, but by his personal seal, of which the bank keeps a recorded impression.

The seals of officials of organizations and provinces are often much more elaborate objects of soapstone or even of jade, several square inches in printing surface, and carved above into intricate

Chinese lions or into other objects respected by the Chinese. Thus it was quite natural for the Chinese at an early period to imprint with one seal a very considerable amount of information. Under these circumstances it is scarcely strange that they came to imprinting the whole page of a book by means of what was essentially a seal carved in wood rather than in stone.

This is the origin of the Chinese block books, which were made like our own of successive paper pages, but with one difference. The Chinese paper was thinner than our paper, and so impressions made on one side were likely to run through onto the other; the recto and verso of a page, instead of being imprinted on the two sides of a single piece of paper, were therefore printed adjacently and then folded into the equivalent of one of our sheets of paper.

One Chinese technique which is closely related to techniques important in modern life, but which has not been given the proper rank in the West that its significance demands, is the technique of well-boring. We are familiar with the drilling of wells for oil, and we know that oil is often found in certain structures in the earth which also contain salt water. Chinese well-drilling was largely for the sake of the salt water in such a formation, and was closely associated with the fact that Imperial China, like royal France, made a large part of its structure of taxation depend on the taxation of salt. The well-drilling was done by driving into the earth tubes of bamboo with cutting bits at their ends. Bamboo, by the way, is a technical material not well known in the West, but quite as important in its own way as is wood. As in our modern drilling rigs, the bit was pounded step by step into the rock to be drilled, being rotated a little at each pounding. With no steam engines or similar devices available, the lifting and dropping of the bit was done by manpower, and this use of manpower involves the employment of a counterpoise so as not to waste the work of lifting the bit each time. The counterpoise took the form of a springy bamboo pole from which the well-drilling rig was suspended.

Another even more important invention which is attributed to China is that of gunpowder. Gunpowder is particularly significant to us now, because some of the problems in its technical employment closely parallel the problems of the use of atomic energy. Very concentrated sources of energy make a maximum demand on our material resources for their control. Thus, when gunpowder was discovered, it was used for producing loud noises and scaring away demons before it was employed in more useful channels.

After that, its employment was primarily military for many centuries. The use of gunpowder in cannons did not come at the very beginning of its military career. In their expeditions of conquest across Asia and Eastern Europe, the Mongols found it a valuable instrument for breaching the walls of fortresses.

From this almost completely uncontrolled use the next step was the cannon, which at the beginning fired roughly rounded stone balls, also for the purpose of breaching city walls. The earliest cannons were quite large devices, and the art of casting metals was scarcely sufficient for constructing siege cannons until somewhat late in the history of the art. Many of the first cannons were longitudinal bundles of rods bound together by forged rings, and were continually subject to the danger of blowing up and of injuring the soldiers who used them. Later on, the art of gun founding was developed, and still later the art of forging and boring even stronger cannons.

The cannon came in earlier than the hand gun, which indeed was at the beginning merely a small portable cannon fastened to the saddle bow of the cavalryman, and fired by a match. The earliest cannons, and the type of all cannons and handguns for many centuries, were muzzle-loading weapons. This was not for lack of attempts to construct breech-loaders of all sizes, but because the early breech-loaders were beyond the technical limits of their time. No good method was found for obturation or, in other words, for the blocking of back-fire between the muzzle and the breechpiece, nor was the metalworking of the time sufficiently reliable to prevent the frequent accident of blowing off the breech. Revolving cannons and machine guns also are exemplified in the early days of the gunsmith's art, but proved to be unattainable in any satisfactory way within the smith's craft as it existed at the time.

In connection with the atomic bomb, we are facing a similar and perhaps insoluble problem of the control of great concentrations of energy. It is for this reason, as well as the accident that the critical stage of atom-splitting occurred just before a great war, that the atomic bomb has preceded the atomic source of controllable power. As far as the uranium and plutonium bombs are concerned, we have parallel to them a technique of atomic piles which may be made to give up their energy at a reasonably slow rate. However, the further extensions of atomic technique are exemplified by the case of the hydrogen bomb, in which the explosion of the bomb as a whole must be preceded by an earlier stage in which a fission bomb is used to bring the final explosion materials to a sufficiently high temperature that the desired reactions may take place. The temperatures needed are comparable with those found in the sun, and are quite inconsistent with the solid state of any form of matter.

Thus the only way in which such reactions can be used is against the inertia of any material as may be in contact with the explosive, so that we depend on the short duration of the explosion to build up effective pressures before the materials around the explosive are dissipated in vapor or in an even more thorough form of dissolution. At the present, there is no reasonable expectation that we can bottle the demons of the hydrogen bomb, yet the history of gunpowder shows that we have previously met demons which seemed uncontrollable, and we have found ways to master and tame them. In fact, the engine that is under the hood of every automobile is nothing but a collection of a number of gun barrels which use a vaporous explosive quite as strong as gunpowder, but confine it to useful purposes. This is only possible because we have at our disposal highly developed metalworking techniques.

I shall have several other cases to recount where the lack of technical aids and a sufficiently high level of the general art and of available materials have held up a valuable invention for considerable periods. As, however, these examples are modern and relatively complex, I shall reserve them for later chapters, in which I can discuss the history of the telephone, of radio, of television, and the like, in a more balanced manner.

The Social Climate and Invention

5

As I have said in the first chapter, when we consider why certain periods have been fruitful in inventions, and why these periods are not always the same as those which have been fruitful in ideas leading up to inventions, we must consider much more than the internal ideology of the process of invention. Let me consider in more detail the example of the Greeks, which we have already discussed in a summary fashion.

The fourth century (B.C.) of Greece was one of the most fertile that the world has ever seen in the basic ideas of science, but it was not a period of particular technical development in the sciences as contrasted with the arts. By contrast, the later Hellenistic period of Greek culture contains such eminently practical names as those of Hero of Alexandria, Archimedes, and Ptolemy. Hero was the inventor, or at least the reputed inventor, of several of the earliest known automata and also conceived and apparently constructed a primitive steam turbine. Archimedes is the founder of hydrostatics. If legend tells us the truth, he contributed a great deal to the defense of Syracuse against the Romans by his ingenious devices. Ptolemy made great contributions both to astronomy and to geography, and appears to have taken an active part in the measurement of the size and the figure of the earth. Why should these two periods, the

Athenian and the Hellenistic, both of which belong to what we consider the ancient classical world, be so different in their relation to invention?

We may look for the cause of this difference by paraphrasing and modifying Plato's statement that in the ideal state, kings must become philosophers and the philosophers, kings. For a great period of invention, the artisans must become philosophers or the philosophers, artisans. There was little or nothing in the training or the orientation of the fourth-century Greek to establish a bridge between the artisan and the philosopher.

Notwithstanding the beauty of the work of the potter and the high level of artistic achievement of artisans in general, the artisan of the great days of Athens, if not a slave, was, more often than not, a humble and peripheral member of society. We can probably make an exception for the painter or the sculptor, but when we have made this exception, we see at once how narrow a group it is to which it pertains, and how little interest there was in those aspects of technical improvement which readily connect with science.

Conversely, the Greek philosopher was a gentleman, participating in affairs of state and in affairs of war, interested in reflection, but with soft hands. The very nature of Greek scientific work bears witness to this. Classical Greek science consists in speculations which can have very little to do with manipulation, and much of it is devoted to the logic of number and quantity rather than to the measure of quantity and computation by means of numbers. The Greek geometric figures were probably drawn in the sand, and Athenian techniques were not implemented, as far as we known, by anything accurate in the way of compasses and rulers.

By an unusual combination of logic and insight, the Greek of the classical age was able to develop such conceptions as that of irrationally related lines and of conic sections which were destined to bear fruit in the physics of two thousand years later, but it would have been easier for him to talk across the centuries with the artificer-scientists of the Renaissance and of the present day than to make himself understood by his own smiths and carpenters.

There had indeed been a time, in the Greece before Greece, when Daedalus and the artificers of Crete had been able to communicate with kings and learned men, but this had become a mere shading memory of the past, and Daedalus is more a contemporary of Watt in his ways of thought than he was of Plato.

We cannot read much of the writings of the Cretans, such as they are, but the plumber civilization of Cnossos bears ample witness to the essential truth of the legend of Daedalus. Why the Minoan, Daedalean civilization arose, and why little more than a memory of it was left in classical times, we can only surmise.

That great flowering of the citizen state, where the gentlemen had the leisure to be philosophers, and the rough work of life was left to slaves and to disenfranchised metics, passed in the catastrophe of the Peloponnesian War as quickly as it had arisen. The later absorption of the city-state in the international super-state of Philip and Alexander completed its extinction. There ensued the Hellenistic period, in which the triumphs of Greek civilization were spread far beyond the compact city-states of their origin, and to a world in which Greek and barbarian met more nearly on equal terms. Alexandria in Egypt was the type of a Hellenistic city, but Syracuse in Sicily was not far behind. In these new states, Greek and Egyptian, Phoenician, Jew, and Syrian, Sicilian and Italic, met as participants in an urban civilization where Greek was the chief language and the intellectual tradition of the great days of Greece set the tone of thought, but where the exclusivism of Greece as a whole, and the Greek city-states in particular, had broken down. In this new world, Alexandria and Syracuse were forerunners of the Paris and New York of the present day. In this new civilization, the artisan and the philosopher had begun to speak the same language, and a limited period of scientific and engineering development was initiated.

We have already seen that the Chinese social system, like the Hellenistic, favored a certain social contact between the artisan and the philosopher, and constituted a moderately favorable climate for invention.

In the later Middle Ages of Europe and during the Renaissance, the social climate again became more favorable for invention and in fact partook in some ways of the earlier Chinese climate. Notwithstanding the power of the feudal lords, the cities were beginning to assert their strength, particularly in Italy and in Flanders. In those cities the guilds of craftsmen asserted much of the authority that the feudal lords asserted on their estates. Even before the full blossoming of the cities into independence and into semi-independence, certain special sorts of craftsmen such as the architect and the adorners of cathedrals had already begun to occupy a high social position.

During the seventeenth century the cities began to lose much of their independent position, although we may perhaps consider the rise of Holland and the period of ascension of Cromwell in England as two great gains made by the craftsmen and the middle class. Nevertheless in France, which was then perhaps the dominant country of Western civilization, the central power of the king and the court came to be asserted more and more.

I am going into these historical matters, not so much for their own sakes, and particularly not so much for the hope of being able to say anything complimentary about them, as because I wish to lay the ground for a discussion of social relations in the present day as they affect invention. The eighteenth century is an interesting period of transition as far as concerns the position of the craftsman, both in England and in France, and represents the beginning of some of the influences we see in the present day. In England a strong Whig ruling class was enabled by the presence of a foreign king to take practically all the wealth and all the authority of the realm. In this grab the position of the bourgeois suffered for more than half of a century even though the aristocracy which had absorbed the church, the schools, the universities, and the army was itself a bourgeois aristocracy. Under the surface, however, the craftsmen were beginning to show signs of consolidating themselves as a new power. I have mentioned Watt already, but he was by no means alone. The late eighteenth century was a period of great road engineers in Great Britain, and we need only to think of the names of McAdam and Tillford. The painter and poet Blake was a craftsman before he was anything else. Indeed, if we go to the satirical paintings of Hogarth, we find favorably impressed upon us in the parallel tales of the industrious and the idle apprentice some of the high status which the craftsman was beginning to win for himself.

Perhaps the highest status reached by a craftsman in the eighteenth century was that of Benjamin Franklin. He was not only the leader of the American intellect but was a honored guest in France and even in England. In France he was both a stimulus and a representative of the new intellectual ferment which was taking place among the Encyclopedists. This intellectual ferment with its classification of the artificer even penetrated to the court, and the favorite pastime of Louis XVI was his work as a locksmith.

This brings us down to the time of the French Revolution, in which we find a very mixed attitude to the scientist-craftsman, at least in those fields of work where he was a little bit removed from the lowest levels of his craft. The men who took Lavoisier to the guillotine are said to have answered to his protests, "The republic has no need of chemists." For all that, I doubt if we can consider that Lavoisier was executed as a chemist, for his unpopular position as farmer of the taxes would have assured a thousand times over that he should become victim of the terror.

Whatever hostility the French Revolution had to the scientist was certainly greatly reduced by the work of Lazare Carnot, almost the only one of the rulers of France during the terror who survived to lead a peaceful and respected life under Napoleon and the new royalty. France knows him as the organizer of victories. In the thin times when all the monarchs of Europe were warring against the

French Revolution, he secured supplies of all military necessities. For example, he had all the old and filthy cellars of Paris scraped to obtain saltpeter for gunpowder out of the efflorescence of the animal matter in the soil. In furnishing scientific help for a country in turmoil and distress, Carnot repeated two hundred years later what Stevinus had done for the new republic of the Netherlands.

The times of the French Revolution and those immediately following were almost as dangerous for men of thought in England as in France. Not long after the chemist Lavoisier was executed by the French terror as a representative of the old regime, that other chemist, Joseph Priestley of England, was driven out of his home by the mob for his sympathies with the French Revolution and was forced to come to America.

I have already mentioned one important American scientist of the eighteenth century, namely, Benjamin Franklin. In the Revolution another American scientist of note took the British side and was the paymaster of Benedict Arnold. He was Benjamin Thompson of Woburn, Massachusetts. During the Revolution Thompson left for England to save his skin, although it must be said that he retained a certain interest in his own country and that the American Academy of Science in Boston was founded by his will. In England he was the founder of the Royal Institute, in which a real attempt was made to bring together the scientific and the industrial side of invention and in which such scientists as Sir Humphry Davy and Michael Faraday found a scope for their activities.

By the way, it will illustrate the close-knit texture of the scientific world at that time, which is so different from the close-knit texture of the scientific world today, if we notice that Benjamin Thompson, after the death of his first wife from Concord (then known as Rumford), New Hampshire, married the widow of Lavoisier and that they lived unhappily ever after. Thompson, for all his keen observation and his ability, was a bit of a climber and ended up as police minister to the Elector of Bavaria under the title Count

Rumford. Here he continued to show his particularly Yankee combination of scientific interest and technical know-how in his study of the boring of cannon as a source of heat. Those of us who now visit the city of Munich will see a visible sign of the life of Rumford in the English garden, the great park of the city, which was established not by an Englishman but by a Massachusetts Yankee.

One of the most interesting social considerations in the sociology of invention is that of the interplay between the craftsman element and the purely scientific element. Perhaps as good a balance as was ever attempted in that field was to be found in the work of Michael Faraday and of his later colleague James Clerk Maxwell. Faraday, who had been laboratory boy for the chemist Sir Humphry Davy, developed a constant theory of electricity on the basis of a language of images and, we may also say, figures of speech. Maxwell, who unlike Faraday was a mathematician and primarily a university man, took these ideas of Faraday and put them into a sharp mathematical language. In doing so, he established the identity of light and electricity and immediately paved the way for the study of Heinrich Hertz's waves and later for the development of wireless and ultimately of radio. It was during this period just before and around the middle of the nineteenth century that the Newtonian physics began to be put at the disposal of the mechanical engineer, and this in particular in Britain. Here we must mention such names as those of Peter Tait and William Thomson. Thomson, who later became Lord Kelvin, is a marvelous example of the synthesis which had begun to appear at this time between theoretical science and industry. Actually he was one of the greatest physicists of his time, but he was also an industrialist of the first order and ended up with a large fortune which he had earned by his services to the cable industry and to other branches of industry dependent on science.

It is impossible for us to think of Lord Kelvin nowadays without thinking of that other scientist, Lord Raleigh. Lord Raleigh was perhaps an even greater scientist than Lord Kelvin, but unlike

Lord Kelvin, his title was a family one and not one that had been given to him for science. The two saw the prestige which scientists had attained during the middle of the nineteenth century and represent an unusually close combination of the intellectual and the practical. This is a period in which it had become inevitable that the career of engineering should develop learning and educational institutions of its own, and even though England was never quite on a level with the continent and America in the new sort of institute of technology, the influence of Kelvin, Raleigh, and others is to be found in both English education and education outside of England. Leaving out the École Polytechnique of France, which was primarily an army school, most of the great technical schools of the world, including the Royal College of Science in London, the Massachusetts Institute of Technology, the Technical Institute of Zurich, and the Technical Institute of Berlin, date to this period around the middle of the nineteenth century.

This was, as I have said, the period of the science lords in England. In the German countries the position of scientists had come to be almost equally elevated, and the careers of some of the German scientists of that time were rather similar to those of the English in their combination of the theoretical and the practical. The name that first comes to my mind here is that of the physiologist, mathematician, and physician Hermann von Helmholtz. Indeed I should mention here the invention of the telegraph by Carl Friedrich Gauss and Wilhelm Weber in Göttingen. As most of these German scientists of high position were titled Geheimrat or privy councillor we may say that the age of the science lords in England is also the age of the Geheimrat.

The Scientific Climate at the Beginning of the Twentieth Century

6

The period of the scientific lords in Britain and the Geheimrats in Germany was one in which the interests of the pure scientist, of the craftsman, and of the industrialist did not come into any sharp conflict. All three of these groups were bent on declaring their independence from the old military and agricultural aristocracy, on the one hand, and from the traditional classical learning, which this aristocracy had considered the only learning fit for a gentleman, on the other. Thus the rise of the technical schools and the policy by which the new Germany after the Franco-Prussian War had decided to industrialize through science were phenomena acceptable to all three.

To these must be added the fact that the last half of the nineteenth century was, par excellence, a period of exploitation in which the accumulated resources of two hundred years of Newtonian physics and of the chemistry that followed it were just being made available to the industrialist; and in which new mines, new forests, and new continents had come to exploitation with very little regard for the fact that they would not last forever, at least in their original newness and profuseness. Thus there was plenty for the scientist, the inventor, and the industrialist to do in concert, without a conflict of interests.

This was the great period of the workshop patent, during which the inventor was conceived as a man who had used some shop, his own or his employer's, to develop a new gadget, and in which this gadget, if it worked at all, was an important discovery which did not need to be reduced to the best possible operation by the techniques of the scientist, in order to bring money into the coffers of the industrialists, and to strike the imagination of the public.

The pure scientist, in so far as the man in the street and the industrialist thought of him at all, was a man of modest living and very little influence, a sort of a poet of industry, to be granted the somewhat pitying indulgence accorded to the poet and the painter, and to be given the freedom necessary to his thought just because he was considered to be of no particular importance.

In the great expansion of American industry which occurred when the Civil War was over and people were free to go back to their peaceful habits of thought and the earning of money, a new element came into the game of invention. This new element was represented by Thomas Alva Edison. Of all inventors, he is perhaps the one who has most struck the imagination of America, and possibly of the world. In his earlier work, he belongs with the shop gadgeteers. For example, such work as he did on multiple telegraphy involves more an ingenious putting together of elements of scientific principles which were already firmly established, than the introduction to science of any new principles. This will also apply to his contributions to the phonograph and to his work on the incandescent light, in which the latter, in particular, involved a pedestrian search through a mass of materials rather than any strikingly new idea. Later on, in connection with the electric light, he discovered what I shall point out later as the greatest scientific innovation ever due to him, and the one invention on which he never made a cent. This was the Edison Effect, according to which, in the presence of a hot filament, the vacuum in an electric light bulb becomes conducting.

However, Edison's greatest invention was not scientific but economic. It was the invention of the industrial scientific laboratory in which a moderately large trained crew of technicians was directed by a central mind towards the making of inventions as an everyday business. In those days when the moral responsibility of the employer was less clearly delimited, and before the unions had won a very considerable territory of their own from the entrepreneur, Edison could be, and was, a tyrannical master. He saw to it very carefully that any invention made in his laboratory was known to the public by the name of Edison, without mention of the particular scientists who may have conceived the idea and to whom its execution was entrusted. He was past master of business and of clever advertising, and he adopted as his personal symbol a certain cracker-barrel workmanlike simplicity of character, which he was not last to exploit.

He was followed, after a considerable interval, by other industrialists and industrial organizations which developed the industrial laboratory far beyond his own level. Among these were the General Electric Company, the Westinghouse complex of industries, and the Bell Telephone Laboratories. But by this time much of the undisciplined exploitation of the Gilded Age was over, and these organizations were ultimately forced to adopt a more enlightened policy as to the recognition of their subordinate inventors and scientists.

By the end of the nineteenth century, as I have already mentioned in the case of Gibbs, science had begun to enter a new phase. Just before the end of the century it was the boast, even though a rather plaintive boast, of the physicists that future generations of science would have nothing left for them but the observation of already known quantities to two or three more decimal points of accuracy. Within the lifetimes of those who had made these predictions, their words were to turn to wormwood in their mouths. The new work of Gibbs, of Planck, and of Einstein was to show that the Newtonian synthesis of science was as relatively inadequate to the new experiments and observations as the Aristotelian synthesis had been in the seventeenth century. A new age was opening in which the scientists had to deal with new phenomena of a character completely indescribable in the old language. Thus, within the first decade of the twentieth century, the scientist of an abstract character, with a personality completely unlike that of the earlier artisan-inventor, was discovering a number of totally new phenomena of which technology would ultimately have to take account.

About this time, the telephone industry, which was founded during the period we have just discussed, and which started as a piece of the gadgeteering which was characteristic of Edison and of the nineteenth-century development of American invention, went through a series of changes which parallel in a more limited field the changes which were taking place in science in general. These changes were remarkable enough to be recounted in detail, and may serve to give to the nonscientific reader an understandable account of the fin-de-siècle change in intellectual climate.

It is only a few times in a century that an industry is confronted with a new need for a change-over which may be vital to its continued existence in any form. Not more than a quarter of a century after the first and highly controversial series of inventions which culminated in the Bell patents, the telephone industry was faced by such a situation.

At the beginning, the very nature of the telephone itself was revolutionary enough to discourage questions as to its ultimate scope. The general early experience of the electrical communication industry concerned itself with apparatus which either worked as a conspicuous success or did not work at all. Thus, once the early telegraphs were devised in any form whatever, it was not long before telegraphy over miles, if not hundreds of miles, became a routine procedure, and was handed over to technicians rather than to scientists.

The first setback of any importance to this sort of hand-to-mouth procedure was the failure of the first transatlantic cable. Nowadays, we know that this cable might well have functioned for years, but that it was burned out at the very inception of its use by what seems to us at the present as the stupid procedure of trying to force its message-carrying capacity by the use of high voltages, which could only have, and did have, the effect of shattering its insulation at some unreachable point at the bottom of the sea.

Apart from this rebuff, early communications research and invention went fairly easily by the discovery of what was originally the gadget device of multiplex telegraphy. As for the telephone, it was conceived in the beginning as a device primarily intended for the convenience of the merchants and professional men within a given city, and never destined to compete with the telegraph in interurban use.

With the arrival of the twentieth century, however, the interurban use of the telephone had become an established practice, and the companies began looking for new fields to conquer. I say the companies, for at that time the unifying connections between different companies even in neighboring parts of the country were not physical. In fact, they did not amount to more than the sharing of a certain patent pool of inventions and, to some extent, of their being financed by a certain group of promoters. Even within one city, rival telephone companies were frequent, and although the mutual separation of their services was a nuisance, it was not regarded as much more than a nuisance.

About 1900 the telephone industry began to take stock of itself and to think in terms of interurban communication as well as of intraurban communication. In the Bell system of telephones, the new company known as the American Telephone and Telegraph Company was formed to exploit these new possibilities. Let us see what the new possibilities were, from a technical point of view.

The very first telephone invented by Bell shared with the contemporary inventions of Amos Dolbear and others the limitation that the energy available at the receiving end was simply a part of the voice energy introduced at the sending end. This is the reason why the electromagnetic induction microphone of Bell lasted so short a time as a practical tool, and why Dolbear's electrostatic induction microphone never came into actual commercial use until the days of modern radio.

Luckily, the young industry found itself confronted in its very early days by the carbon microphone, which is an instrument of a very different nature. In the carbon microphone, the energy that goes through the system and that acts at the receiving end is inserted into the system by batteries. The currents of these batteries are modulated by the variable resistance due to the variable pressure of the air caused by the vibrations of speech, and transmitted to a pile of carbon granules. This system can, in fact, transfer to its output much more speech energy than was given to it originally by the voice. In short, the carbon microphone is a variable resistance that acts as a powerful amplifier. It was the amplifying effect that led to its introduction, even though amplification was then so isolated a phenomenon that there was no general theory for it. Thus a good angel brought gifts at the birth of the telephone, although the extraordinary nature of these gifts was not appreciated until much later.

The local use of the telephone did not suffer at the beginning from a scarcity of power, nor, in fact, was the scarcity of power a critical factor in limiting its use over long distances. Much before speech put into the telephone would cease to be heard at a distance by the mere weakening of its energy, it would become an incomprehensible gibberish. That is, distortion and not attenuation was the first obstacle that had to be overcome by intercity systems.

In many countries people had begun to discuss the engineering and mathematical problems of distortion, and in none more actively than in England. Not every country has been so devoted to private enterprise as the United States. By the end of the 1880s, the British telephone system was being taken over by the Post Office. The chief public servant in charge of the technical end of the Post Office was an energetic but none too intelligent engineer by the name of William Henry Preece, who was later knighted.

Preece had the idea that the difficulties of long-distance communication to which the telephone was manifestly subject were due to the fact that electric circuits as used in his day were defective in the property known as capacity. Capacity enables a conductor to take in a considerable charge of electricity with a small rise in potential. There is a verbal similarity between the concept of electrostatic capacity and the conception of of message-carrying capacity, and this bad pun is tempting to shallow thinkers.

As a matter of fact, capacity, instead of being the friend of the communication engineer, is his grossest enemy. Indeed, the transatlantic cables, including the early one which had failed, were gigantic Leyden jars of a fabulous capacity, and had to have a large quantity of electricity poured into them at one end before they could be filled up and able to disgorge this electricity at the other end. It was ignorance of this fact that had led to the debacle of the first transatlantic cable.

This was understood by William Thomson of England, later to become the Lord Kelvin who appears in an earlier chapter of this book. He saw that because of this great capacity, the receiving instruments of the cable system would have to be of such a nature that they would catch the first outpouring of the cable, before the rise of potential had reached any considerable size. He did this with the aid of sensitive siphon galvanometers, which were the mainstay of cable instrumentation for more than a generation.

Thomson could not have been very pleased with the ignorant but authoritarian outpourings of Preece, but it was not Thomson who was the gadfly who made Preece's existence miserable. This was an undersized, hungry, deaf, cantankerous little electrician by the name of Oliver Heaviside, who frequented the meetings of the

Institution of Electrical Engineers, and who sent memoranda to *The Electrician*, written in a Swiftian style. These were so effective that a reprimand from Heaviside had the vesicating character of a horse-whipping.

Heaviside was born poor, lived poor, and died poor; he was sincere, courageous, and incorruptible. In addition, he could use the very limited mathematics available to the electrical engineers of his day with an unorthodox skill that was to disconcert a whole generation of mathematicians.

Heaviside attacked the problem of distortion at its very roots. He was not the only person to have seen the problem, but he was the first person to think of it sharply and habitually in engineering terms which were relevant to the new technology.

According to him, for a circuit to be nondistorting, a certain very sharp balance between four quantities was necessary. These were the resistivity of the line, the leakage to ground, the electrostatic capacity, and a certain quantity of the nature of inertia known as inductance. While an ordinary straight line has inductance, this is greatly increased when the straight line is replaced by the winding of an electromagnet.

Heaviside asserted that the ordinary communication line had too much capacity, not too little (as Preece had said), and that the way of improvement lay through the introduction of more inductance along the line. Preece blustered urbanely, but this gadfly, this wasp, was not afraid of any bluster. With the aid of his brother, who was an engineer in the Post Office, Heaviside smuggled an out-of-hours employment of one of the long lines of the Post Office to test his theories. This test was not conclusive, but it was not performed under conditions sufficiently satisfactory to refute Heaviside's confidence in his own ideas.

These ideas were published first in *The Electrician*. Later many of these *Electrician* papers were collected in a series of books published by Appleton. Appleton regretted the publication, for the books did not sell and were finally remaindered. A generation later, the original books and at least three pirated editions became necessary constituents of the library of every communication engineer. One of these pirated editions was printed in China.

It gradually came to be recognized that Heaviside was right, and that the distortionless line was the first step to long-distance telephony. By this time, however, ten or fifteen years had passed, much more than the period in which patent proceedings could be initiated. The invention had been dedicated to the public, which means that it did not exist as a commercial invention, and that no rights in it could be acquired by any method whatever. It was on this invention that the newly founded American Telephone and Telegraph Company (AT&T) sought to base the new technique of long lines. There was no way to acquire property rights in the long lines idea that was not equally available to any other entrepreneur who wished to start work in this direction.

It must be remembered that the various alternative techniques to the distortionless line which have since developed were not available to the founders of AT&T. I have said already that the carbon microphone was intrinsically an amplifier. Certain of the early lines of AT&T actually used it in this role by incorporating repeating stations in which a telephone receiver was apposed to a carbon microphone, thus giving new strength to the enfeebled message. However, this trick was satisfactory neither from the point of view of the amount of power amplified nor from the point of view of dependability. It was merely the best thing that could be done at the time, and represents an intelligent makeshift which spanned the period between the first development of long lines and the discovery of the vacuum tube amplifier as a powerful and dependable device.

The vacuum tube amplifier has a most anomalous position in the history of invention. It emerged from a complicated patent situation in which counterclaims were held by Lee De Forest and Sir John Fleming, but the original discovery emanated from neither of them. It was an application of the Edison Effect, of which I have already spoken. Edison, as I have said, found that a current could flow inside of the bulb of an electric light between a hot filament and another electrode when it could not flow if the filament were cold.

From the commercial point of view, the discovery was made too early. The Hertzian waves had just been discovered and were very far from any state in which they could be supposed to be useful for wireless telegraphy. Thus Edison's invention, like the work of Heaviside, had lapsed as a salable patent before the outside development of the art made it of any consequence, and it served merely to confuse the legal struggles which arose at a later time when vacuum tubes were found to have an obvious significance for wireless telegraphy.

To go back to Heaviside, it was an utter necessity for those who were going to develop a system of long lines to possess some property rights that could plausibly be interpreted as covering the Heaviside invention of the distortionless line. Without such rights, they would certainly have been unable to be sufficiently secure in their sole exploitation of the new ideas to protect the heavy and risky gamble on which they had entered. They might even have found themselves on Queer Street if somebody else had made a fundamental invention that might—for such things have happened—deprive them of any legal claims to exploit their own ideas. What happened is interesting from both a technical and a moral point of view, although I shall leave much of the moral interpretation of the facts to the reader.

Naturally, the telephone company went full speed ahead to develop the Heaviside ideas and to see whether there was any point which Heaviside had missed which could be made the basis of new patent claims. In doing this they made use both of their own man, Campbell, and of outside scientists. The chief among the latter was Michael Idvorsky Pupin. Pupin's book, *From Immigrant to Inventor,* was for a period almost the bible of the American schoolboy eager to succeed in invention.

The main object of both investigators, Campbell and Pupin, was to find some respect in which Heaviside's work, if not wrong, was incomplete. Heaviside had developed his theory in the first instance for a continuously loaded line, but he was quite aware of the fact that the new inductances to be used to correct the distortion in the line would almost certainly have to be distributed at intervals. He suggested an interval something like one per mile, which, in fact, would not have given too bad a result.

He nowhere stated explicitly the principles on which the widest admissible spacing of the loaded coils might be determined. Both Campbell and Pupin went after this lack of explicitness in Heaviside's theory, and both of them found that if the loading coils were given a certain density of distribution along the line, then for frequencies lower than a certain level, the line would behave essentially like the uniformly loaded line. For higher frequencies, on the other hand, the oscillations would get tangled up by the successive reflections at the loading coils, and would not get through.

Probably Campbell's work was in advance of Pupin's, but since Campbell was already an employee of AT&T, there was a distinct possibility that a patent by him would lack the convincing power on a court of a patent purchased from an independent inventor. It was as a result of this that AT&T gave Pupin half a million dollars for his various patents. Besides the spacing of coils, these contained various claims concerning the technical method of constructing the toroidal coils already described by Heaviside.

Both Pupin and Campbell went further with their work. In their later developments, the work of Campbell was clearly the more thorough and the more comprehensive. One of the tricky possibilities contemplated by the two inventors independently was that of converting the disadvantages of the long line with lump loading into advantages for different purposes. It was foreseen that there might be cases in which the frequency cutoff point of the loaded line might be an advantage rather than a disadvantage. This led to the development of the theory of the wave filter, which is able to pass messages of one frequency range, while destroying messages of another frequency range. For many years, indeed, the design of these wave filters bore distinct traces of the long line theory from which they had in fact emanated. Later developments of the wave filter have gone far to eliminate the traces of the artificial loaded line.

With these patents the Bell organization had achieved its main end, and there only remained a question of sweeping up the pieces. The Bell engineers felt a very considerable hostility towards Pupin. They believed that he had superseded, not only Heaviside's claims, but also the rightful claims of Campbell. However, this was a family scandal which could be and was hushed up, as far as outsiders were concerned.

There remained the much more fundamental question of what to do with Heaviside. On the one hand, he might have been able to put up some claims which, weak as they were in his own private hands, might yet have become ominous if some strong rival were to try to derive rights from him. On the other hand (at least I like to think so), the Bell people themselves, who, after all, were engineers and had a considerable justified respect for Heaviside's work, did not want to see him go altogether empty-handed. What was a trivial sum to the company was, they very well knew, a matter of life and death, of eating and starvation, for poor Heaviside, whose increasing deafness and crankiness of temper were making it harder and harder for him to secure new employment. The Bell people offered him a sum for his rights, and it might very well have been that this sum would have proved appreciable from his point of view.

Heaviside turned the offer down flatly, declining to accept one cent unless the company were to give him full recognition as the original and sole inventor of the loaded line. This the company could not do, for if it had, the whole investment of half a million dollars which it had made in Pupin would have ceased to be of any use.

I do not wish to convey the idea that Heaviside would have been easy to deal with under the best of circumstances. Like many deaf men, he was a strong individualist, and he was as bristly as a porcupine. Those who tried to help him at difficult periods found him no very mild recipient of favors. I have talked with B. A. Behrend, who went down to Heaviside's little cottage at Torquay to persuade him to accept an honor conveyed by the American Institute of Electrical Engineers. Heaviside was recalcitrant. It was only because he was made to realize that Behrend, his friend, would have been very much hurt by his refusal, that he was persuaded to accept the honor. The officials of the British Institution of Electrical Engineers had already had a very similar experience.

Heaviside was not precisely amiable, but he was fair. In fact, he had made the great invention on which the future of the Bell Telephone system depended. If he was not able to be rewarded for his merits by an adequate provision for his old age and by due credit for his invention, he was not going to be a party to a lie. He had the grim satisfaction that in his own way he was causing the officials of the telephone company more worry than they were able to cause him. A man who in all essential respects has foresworn the world and accepted poverty is invulnerable.

Corporations have no feelings to be hurt, nor does a man of easily moved feelings rise to high office in one, but there was a man who was hurt and hurt greatly. While Heaviside had the sardonic satisfaction of incorruptibility, Pupin was in a really difficult situation. He had accepted half a million dollars which was legally his for an invention which he had only made in a very Pickwickian sense.

This half million had already passed, and even if Pupin wished to return it, which I very much doubt, I cannot see any way in which he could have done so. Thus, whether he wished it or no, he had to become a supporter of the deal by which he was enriched. However, his ego needed more and more fortification as time went on, and in his books he began to aggrandize his own role and to belittle that of Heaviside. This may not have been noble, but it was very, very human. Furthermore, as I have said, in the Bell Company itself, whatever the officials may have thought of the coup of gamesmanship by which they had acquired a legal command of the situation, the engineers were not very friendly to Pupin. Thus the shoe began to pinch Pupin harder and harder, and he was not a big enough man to endure this queer combination of commercial triumph and moral bankruptcy without protesting more and more. For those who read between the lines, *From Immigrant to Inventor* is not a typical American success story, but a cry from Hell.

The fact is that circumstances had joined the story of Prometheus with the story of Marlowe's Dr. Faustus. Heaviside may have been a very snuffy lower-middle-class Prometheus, but at least he had snatched a piece of fire for mankind. If the vultures of poverty and the sense of persecution were gnawing at his liver, he shared with Prometheus the sense of having performed a godlike feat. Pupin, on the other hand, had wrapped his soul as part and parcel of a commercial bargain. When a soul is bought by anyone, the devil is the ultimate consumer. Even a public penance was denied Pupin. Although he was unable to contain himself in silence, the lies and bluster to which he was forced to resort must have echoed hollowly in the empty space where his soul had been.

The Present Social Environment of Invention: Megabuck Science

7

As far as the first two decades of the present century are concerned, the new and developing non-Newtonian physics had not yet had much chance to infect the streams of invention before the beginning of World War I. The science of the war itself was of course devoted to the technology of more immediate necessities. Actually, most of the main new elements in the technology of this period go back to the science of the third quarter of the nineteenth century. As I have pointed out, the telephone began to make new requirements on the mathematics of electric circuits. At the same time, Maxwell's discovery of the quantitative identity of light and electricity, which had already found an application in the waves of Hertz, came into full use in the wireless of Guglielmo Marconi.

The vacuum tube, the direct descendent of the Edison Effect, was only just beginning to come into its own during the war. The proper technical use of the electric motor, and in particular of the fractional horsepower electric motor, was barely beginning to be seen. The finer electric technique of radar was not discovered until just before World War II, and the new phenomenon of radioactivity and even the somewhat older phenomenon of the X-ray were still in a more or less rudimentary stage of application. It was World War II, during which the nuclear studies of the previous decade led to the development of the atomic bomb, in which the technology was for the first time a technology based on thoroughly twentieth-century science.

During all this period, however, the research laboratory had become more and more the general source of new inventions, not only in America, but also in Europe. It was in Germany, in fact, that the chemical side of the industrial research laboratory had first come into its own. Moreover, not only was the research laboratory being more and more used, but the very conception of its being had changed from that of the small-scale laboratory to that of what we may call the factory laboratory.

Several converging factors played a role in the emergence of the laboratory big enough to be a factory in its own right. The work of Kammerlingh Onnes, in Holland, on the liquefaction of gases, had required an establishment of unprecedented size. (I should note that since Onnes's day the progress of technique has made it possible to duplicate much of his work within relatively small dimensions.) The Bell Telephone Laboratories and the research organizations of the Westinghouse companies and General Electric in this country, as well as of the Siemens combine in Germany, had attained a considerable size, not only in the all-over space and number of men allotted to them, but in the bulk of the apparatus used as well. Pyotr Kapitsa, in the 1930s, first in England and then as a semivoluntary captive in his native country of Russia, had carried out experiments on magnetism involving the sudden short-circuiting of heavy machinery, and he was one of the first to approach the scale of experimentation used in the United States during World War II in Project Manhattan and the development of the atomic bomb.

By this time, really modern science had begun to be used sparingly for the purposes of invention. At the Bell Telephone Laboratories, Clinton Davisson and Lester Germer had verified Heisenberg's predictions of 1925 concerning the wave character of the electron. About 1921, Sir John Cockcroft and Ernest Walton had taken the first steps towards atom-splitting: that is, towards the reduction of the uncontrollable atomic disintegration which had been discovered by the Curies to a controllable process. By the time at which we were ready to explode the atomic bomb, it was a foregone conclusion that the invention of the future world would depend in a large measure on new science, and even on science as yet to be discovered.

This change must have caused consternation on the part of the industrialists, as soon as they became aware of it, even as it caused consternation on the part of the man in the street, to the extent to which he was allowed to become aware of the new climate. No longer was the inventor to be the man of the shop, with whom the industrialist had long been accustomed to deal. No longer was the pure scientist, with his incomprehensible notions and notations, to remain the unworldly, harmless person whom the industrialist had taken him to be. On the contrary, he had become the repository of great and possibly destructive powers. If he were to insist on controlling those powers that he alone understood, he might well become a curb on the all-embracing control of public affairs which the industrialist had by now taken to be his natural right.

Until the period between the two world wars, science in general had offered rather slim pickings to the ambitious young man on the make. There had indeed been exceptions. The gold-making alchemists at the court of the Emperor Rudolph may be regarded as the prototype of the megabuck scientist of the present day, with their demands for more and more money to be spent on their projects. Cardano is as typical a specimen of the adventurer of the Renaissance as was Cellini, albeit Cellini's adventures were in an artistic milieu and Cardano's were in a scientific milieu. Even the great Leibniz had a keen eye for the main chance.

It is not, however, until we come to our own Yankee Ben Thompson of Woburn that we find almost ready-made the prototype of the scientific adventurers who have beset the present age of great projects. Where the carcass is, you will find the flies buzzing, and where the money is, you will find the adventurers, or, to use a more living word, the racketeers.

Benjamin Thompson's racket was first to make himself solid with the English aristocratic government in America, as represented by Governor Wentworth of New Hampshire. Later, when he had to flee to England, he found his spiritual home in the patronage of the British aristocracy, and still later in that of the Elector of Bavaria. Since his time, there has been a bear market on kings and aristocrats, and a bull market in Big Business. Not until the period just preceding World War I, and the period between the two world wars, did the career of the scientist become tempting enough for those gentlemen who make their fortune by their wits.

After all, the career of an ascetic has at last one advantage, that nobody is tempted to batten very much on his worldly goods. However, the present century was not very far along before it became clear that business, industry, and engineering had a back door which was not locked, and that offered many temptations for the experimentally minded. The role of the scientist in industry was ever increasing, and was beginning to offer an entrée into the circles of power for those gentlemen who had nothing to invest but their wits. I have already given an example of the adventurer-scientist in Michael Idvorsky Pupin.

Here I am forced to draw conclusions concerning the psychology of the young scientist on the make which I may not be able to justify in any specific case, but which, in my opinion, stand out very clearly if one observes the situation as a whole.

These young men, for all of them were young once, seem to have a lack of trust in the wings of the spirit and, indeed, a general doubt of the existence of anything which one can properly call an ideal. Though they have no spiritual wings at all, they seem to be under a deep necessity to get there just the same.

Others of these young men on the make started out with a drive to further science for its own sake, and within a limited range, there were those of them who were not poor in ideas. However, not a few came from origins which, if not origins of grinding poverty, contained at least enough elements of poverty around them to make them determined that *that*, at any rate, was not an evil to which they would readily submit.

The only investment they had to offer was their brains, and they made quite sure from the beginning that this investment would pay fat dividends.

In more than one of them there was a stage in their careers during which they might have turned in either direction—to a true devotion to science as science, or to power and that minted symbol of power, money. They found, however, that Mammon is a jealous god. I suspect that in more than one case, the latter-day Benjamin Thompsons made a cool evaluation of their own abilities and came to the conclusion that the top rank in science, as an affair of devotion and of the intellect, was not for them. I further suspect that in more than one instance, this evaluation was supplemented by a deep sense of inferiority going back to the days of their youth.

At any rate, these bright young men saw the big laboratory coming and decided that it was exactly their meat. They had a well-founded fear of the prowess of other scientists and could never feel secure among any group of scholars that was not rendered impotent by the subdivision of their task and by the closure to more active intelligences of adequate avenues of information.

I do not suggest that this opposition to the old-fashioned scientist was fully on a conscious level. The up-and-coming boys believed what they wanted to believe, that the new technology of the Edison laboratory had replaced the individual intelligence and had rendered individual devotion no longer necessary. They frankly admired the machine that replaced men and the mass attack that blotted out men into anonymity.

In order to achieve their ends these prophets of the new day of tied-to-industry science and of mass science had to capture the existing organization of science for their own ends. Probably they rationalized this capture as the facilitation of the wave of the future. For all that, the methods used were not unlike those of the unsainted Alphonse Capone.

In the organization of science, certain societies and academies which shared the two purposes of technical advice to the government in times of emergency and the recognition of achievement were important seats of dignity and authority. In the times of peace following the Civil War these organizations had found less and less to do with the briefing of the government. They tended more and more to become societies of recognition and bulwarks of vested interests of arrived scholars and scholar-administrators. During World War I, and even more during World War II, they failed to resume their governmental function and were supplanted by a series of new organizations, assembled ad hoc. The existence of these honorific organizations raises a certain interesting group of questions concerning honor and honors in science.

The very age and dignity of the scientists who had arrived and their increasing age with the continued development of our academies tended more and more to deprive these societies of their declared function of encouraging science and scientists. With their limited numbers, which were not increased by any means as rapidly as the growth of scientific activity in the United States, the period at which they recognized a man was bound to come later and later in his career, as generations succeeded one another.

Now a man who has not attained a certain measure of recognition somewhere around the middle of his career is very unpromising material for late recognition. The result is that these societies tended to confine themselves more and more to the recognition of those who had already been recognized. In other words, membership in an academy became more and more a secondary criterion of merit. Thus the honorific societies, like the institution of honorary degrees, gradually became devices for the protection of vested interests, for the recognition of recognition itself, and ceased to perform a function of really original value.

Among the vested interests of science in this country there were certain academic interests which had already received their full mede of recognition, and there were the newer industrial laboratories clamoring at the gate. The fact of the real underrecognition of those in industrial laboratories furnished a partial justification for an attempt to give these institutions a larger share in the academies, and offered a certain degree of hope for the new industrial politicians of science. These were trying not simply to open the doors to what had previously been considered a lower caste in science, but actually to turn the academies over lock, stock, and barrel to the industries. I will not go into the methods by which this was accomplished. The process ended in the unqualified glorification of big, expensive, and highly organized science, of the sort suitable to the industrial laboratory, at the cost of the individual thinker.

This was a period in which there was a real and justified growth of ideas of automation. By this I mean both the taking over of certain human functions by the machine, and their transfer to certain mechanically organized great undertakings, in which the use made of flesh and blood workers may have been said to follow the development of the machines and to have imitated them.

This new step, though it contained certain justifiable elements, was carried out largely by those who had a peculiar intellectual and spiritual preference for machine ways of thought. They hoped, and I do not think that they always hoped unconsciously, that the new machines, that the big organization, would replace the need for individual thoughtfulness, and would remove the continual threat of the free-lance scientist and the man of small demands against the supremacy of established position and authority. In accomplishing this, they used a considerable amount of double talk which I think it is necessary for us to analyze.

One of the great problems of modern science is that of the sheer bulk of publications. There is a considerable quantity of fingering through the leaves of existing publications, which has become oppressive merely because of its size. This had led, not only to the compression of libraries by microphotographic methods, but to the development of instrumental ways of exploring these libraries and of cataloguing them.

Such instruments have a value, but this value is strictly limited. There are connections between ideas in different fields which can only become apparent to those who have worked in more than one of those fields, and which must escape the routine human library cataloguers and even more the mechanical cataloguer. I have already mentioned a field in which epidemiology and the study of the properties of combustible gases meet and join one another. Thus papers in one field may be relevant to work in another. Once this relevance is discovered, it may be introduced into the handbook of the cataloguer, or the taping of a cataloguing machine, but I cannot conceive of any process by which the idea of such a relationship would occur to a person who has not done some individual work in both fields.

Such a person would indeed be greatly aided by the proper use of a cataloguing machine. However, I believe, and I believe with serious grounds, that many of those who are making wide propaganda for the cataloguing machine have in the back of their minds an expectation that the cataloguing machine will not merely supplement but will replace the work of the individual seeker of cross-connections. I believe that the propagandists for mechanical aids in cataloguing are making a considerable and not altogether unsuccessful effort to establish this non sequitur in the minds of the public.

Machines for translating and, in particular, machines for translating speech into writing, and writing into speech, are conceivable, and we know the general way of constructing them. They have been promoted by much the same group of people who have been interested in cataloguing machines. They are subject to many difficulties; in fact, their construction involves all the difficulties that we encounter in the nonmechanical problems of translation. We all know the story of the English-speaking person who translated from the German a saying, "The ghost wants to, but the meat is rare." This is quite possible as a literal translation of the sentence, "Der Geist will es, aber der Fleisch ist schwach." This of course means, "The Spirit is willing, but the flesh is weak." How are we to make sure that a machine will not make this stupid mistake?

To brief the machine, which is to secure itself not only against this mistake, but against all mistakes of a comparable degree of ridiculousness and plausibility, would have to involve almost infinite labor and subtlety in its construction. Since it is supposed that the machine is to be used by a scientific translator who demands accuracy, every now and then he is likely to run into a gaffe and to be frustrated. This is not such a serious matter when the gaffe is as ridiculous as the one I have mentioned, but there are borderline cases when both conceivable translations will have a meaning, and which are just enough different to pass a casual inspection. What is the result of this as to the effectiveness and field of use of a translating machine?

The answer is that such a machine in the hands of an expert translator and linguist may greatly reduce the effort he must make in translating a language in which he knows something, but with which he is not perfectly acquainted. When he comes to a difficulty of this sort, it will flash a danger signal in his mind, and he then must be prepared to make the usual effort of a translator unhelped and unencumbered by a translating machine. If he is a really first-rate linguist with a wide acquaintance with a considerable scope of languages, he may be able to guess what is likely to be wrong, even in a language of which he knows very little. For such a man, a translating machine may speed up his activity and increase his usefulness by quite a considerable factor.

However, in the hands of a man without a linguistic education, the machine is worse than useless—it is positively dangerous. A silly translation of a phrase in a diplomatic document might easily cause a war.

In the hands of the expert translator, the machine which speeds him up is just as useful as his ability to be speeded up permits. If he is already working at top intellectual capacity, it is of no use at all. To match the high-speed translating machine, we must use translators of abnormally great activity, rather than translators who need to use a machine as a crutch. Thus the effect of the machine of this sort in intellectual work may be to decrease the number of first-class brains needed but to increase, and greatly increase, the demands made on these brains.

What is true of cataloguing machines and of translating machines is true (under proper rephrasing) of computing machines. To be able to stand the speedup of such a machine, a computer must have a better and not a worse knowledge of his mathematics. It is only when first-rate tools are put in the hands of first-rate men that we avoid an expensive waste of effort. If we are indifferent to the needs of avoiding such expense, we allow what I may say is a criminal waste of effort.

I have been made acquainted with projects in which hundreds of thousands of dollars were asked of the government to make a recording apparatus capable of making over fifty records at the same time. It is a difficult technique to combine fifty sequences of data in such a manner as not to throw away far more informative material than one saves in the final result. It requires a high degree of mathematical sophistication.

When this request for funds for such a machine is made by a man with not even the mathematical equipment of a college senior specializing in mathematics, he is either stupid or criminally indifferent to his waste of government money. To my mind, a man who had made such a demand on government funds and who is not exonerated on the ground of his childish stupidity and ignorance is guilty of an attempt to defraud the United States, and if not punished, should at least be sent to Coventry by all decent scientists.

Be this as it may, the present desire for the mechanical replacement of the human mind has its sharp limits. Where the task done by an individual is narrowly and sharply understood, it is not too difficult to find a fairly adequate replacement either by a purely mechanical device or by an organization in which human minds are put together as if they were cogs in such a device.

However, the use of the human mind for evolving really new thoughts is a new phenomenon each time. To expect to obtain new ideas of real significance by the multiplication of low-grade human activity and by the fortuitous rearrangement of existing ideas without the leadership of a first-rate mind in the selection of these ideas is another form of the fallacy of the monkeys and the typewriter, which already appears with a slightly simpler statement in Swift's *Voyage to Laputa.*

Let us grant that if the monkeys keep tapping the keys at random for a sufficiently long period they will ultimately write every book in our literature. The difficulties are two: First, that the period in question will be vastly greater than the entire contemplated time of the existence of the universe, and, second, that after the monkeys have typed out, let us say, Shakespeare and the Bible, they will have tapped out a much vaster body of nonsense from which we must extract Shakespeare and the Bible.

This mass of sense and nonsense, predominantly nonsense, will only contain Shakespeare and the Bible in the sense in which the marble block of the sculptor will contain his work of art. In other words, the work of the monkeys, who constitute merely a paraphrase for many of the scientists working in the large modern laboratories, will not be of the slightest value unless it is put in the hands of a superb intellect who will not find the mass work of the apes of the slightest assistance to him. It will be far better for him to forget an infinity of monkeyshines and apply his mind directly to the problem before him.

The Present Social Environment of Invention: Megabuck Science, Part II

8

It will appear from what I have said already that I consider that the leaders of the present trend from individualistic research to controlled industrial research are dominated, or at least seriously touched, by a distrust in the individual which often amounts to a distrust in the human. I wish to give more of the details and of the particular manifestations of this antihuman trend before I comment in detail on what it has accomplished and on its dangers.

The great laboratory of the present day, whether it is an organization of a corporation or an agency of the government, whether it is found in the United States or in Soviet Russia, whether its putative (or even its actual) purpose is to earn dividends or to fulfill the demands of a bureaucracy, is generally devoted to the accomplishment of a specific task. This task is mapped out by some sort of planning board, and is subdivided into subordinate tasks, each within the realm of some specialist. These specialists are primarily hired for their competence in certain limited fields, outside of which they are not encouraged to go, or even to satisfy their curiosity. This is partly to prevent them from wasting their time. In a large project, time is likely to amount to quite a considerable amount of money, and the slightest lack of organization will not only eat up the time of the individual worker, but will put the entire organization at sixes and sevens.

Moreover, the secrecy maintained by such an organization without and within is largely a matter of fear. In the case of the government, this fear is primarily that of treason, and of endowing a potential or actual enemy with possible weapons. In the case of an industrial enterprise, the competition is the enemy, and it occupies a position quite analogous to that of the foreign government for government undertakings.

Besides this external fear, there is an even more immediate internal fear. The morality of a struggle to the death which obtains between companies also obtains between officials and employees in any one company. For a superior official who is forced by the very nature of his work to employ subordinates of a very considerable intellectual standing, this fear is intense. He can only allay it by hoping to be able to employ subordinates of the very lowest intellectual standing which is sufficient for the work that is to be done, to build up the weak employee of barely admissible qualifications by giving him the maximum of inhuman, noncompetitive machine aids, and by reducing the stature of his juniors by denying them such information as may enable them to see the picture of the work as a whole and to emerge as his potential equals.

There are indeed occasions on which this technique of minutely subdivided piece work is effective, despite the deeply cynical attitude implied by it. In a war, when the greater part of the ideas are already available to spark a vast project, and when the main difficulties to be overcome are primarily those of bulk and of engineering development, something of the sort is clearly indicated. However, the big project is, I believe, limited in its field of applicability. However the advocates of the machine in steel and the machine in the flesh may feel, it will not do as the basic pattern of all future research and invention.

In the first place, it presupposes that the work of invention take a form assignable in advance, and very generally the form of the solution of a specific preassigned problem or a group of problems.

The normal process is to assign such a problem to a team of scientists in various related fields whose task it is to solve the problem. In view of what I have already said, this represents the exaltation of an important part of the work of invention into the whole of it.

At the stage at which it is possible to specify one's task in advance and to give a fair account of the various talents which are needed to accomplish it, the task is already very well along. The earlier stages of scientific discovery are made at a level at which the particular task to be solved is as yet undetermined. Thus they do not lend themselves well either to the great industrial undertakings or to the cost accounting which has become so important and even dominant an element of all such undertakings.

This means that the great industrial laboratory is not the suitable place for the earlier exploratory work, and that much of it had better be left to the private scientist or to the university laboratory. Even if these take a very serious view of their final obligation to society, this obligation is not hampered by the need of giving a day-to-day account of the precise extent to which they have succeeded in discharging their obligations.

Besides the general advancement of science on all fronts and the building up of our ideas to the point at which they are ready to ripen into action, there is one very special technique of invention which can indeed be followed in large laboratories, and which to a certain extent is already so followed, but which is more remote from their usual proceedings than the search after the accomplishment of highly specific goals.

It is what we may call the inverse process of invention. At many stages we possess new constructible tools or new intellectual tools which obviously are bound to increase our powers considerably in some direction or other. The question is, in what direction? It is just as truly a work of invention or discovery to find out what we are able to accomplish by the use of these new tools as it is to search for the tools which will make possible a specific new device or method.

It is not the exception but the rule for new tools to be undervalued or at least misvalued. In the early days of the electric motor, the chief use seen for it was the transmission of power from a central generating station to plants which, up to that time, had produced their own power by steam engines or hydraulic turbines. The motor was considered as a sort of secondary equivalent of a prime mover. When the steam engine or water wheel was discontinued, a natural thing was to use the investment already made in belts and pulleys by hitching up exactly the existing sort of a factory to a large motor; and the only change was that the power used was bought as electricity from a utility company instead of as coal or water rights.

It took a considerable time for engineers and industrialists to see that this use of electric power was not intrinsic in the electric motor itself, but rather belonged to the particular economic and technical stage in which these factories found themselves, with a large investment in the pulleys and belts and shafts of the millwright, and in machinery which was destined to be operated from these belts and pulleys and shafts.

Under these conditions, the discovery that the individual machines in a factory could be advantageously operated each from its own motor, which might well be a fractional horsepower motor, was a true invention. The question that this act of invention answered was not, How can I run such-and-such an existing factory most effectively? but rather, What is the real meaning and the appropriate function of the electric motor itself? The gap between the first electrification of industry and the rise of the fractional horsepower motor to its own was at least forty or fifty years. There was no reason why it might not have been made at any time after the original invention of the electric motor, by an engineer curious enough to wish to see what the invention of the motor actually meant. I doubt if this profound development can be attributed to any one man, and I am inclined to believe that it came by a simultaneous awareness, in many independent places, of the appropriate function of the electric motor.

The present example will show that the social and economic importance of the inverse invention will compare with that of the direct invention. The social consequences of the discovery of the proper function of the small motor have proved to be enormous. In the first place, a whole art, the art of the millwright, has become obsolete. It had not been a simple matter to establish long lines of shafting, with their pulleys and belts, so that they could be properly lubricated, so that their alignment could be mitigated by universal joints and parallel couplings.

Once the shed factory of the millwright had been established with its shafting and belting, it was not an easy task to rearrange the machines used. The factory was forced into a Procrustean bed which imposed upon it a conservatism in every respect. It was a dangerous place, where belts and pulleys could seize the unwary worker and whirl him about to his death. It was a dirty place, where drops of oil were being spattered everywhere from inaccessible oil fixtures up near the ceiling. It was a vast cavern, difficult to heat and difficult to light, where naked electric light bulbs concealed by their glare almost as much as they exhibited. In short, it was one of the crueler circles of the hell of the Industrial Revolution.

With the coming of the fractional horsepower motor, integrated into the structure of each mechanical tool, the need for this situation disappeared at once. Belts could be made short or even replaced by safe trains of gearing working under adequate and designable covers. The arrangement of machines upon the floor was no longer subject to any problem of alignment.

The machines could be clean, at least as far as the spattering of oil was concerned, and it became no more difficult to light them than the desks in a clerical office. The lighting could be adapted to the visual needs of the worker, and as the mercury vapor lamp and later the fluorescent lamp came to their own, they could be employed at precisely the most advantageous points. The factory room itself was no longer a cave of the winds with its walls pierced again and again for shafting and belting, but became a closed space easily heated and easily ventilated. All of this came as the answer, not to a question, How shall I build such a factory? but ultimately to the deeper question, What is a reasonable way to build factories in view of the availability of the fractional horsepower motor?

The social and economic consequences of the fractional horsepower motor go far beyond this. Every country garage can buy its machine tools with built-in power. Every amateur of woodworking and furniture making can have his shop in his cellar. The motors have become so small that they can be fitted into semiportable tools such as saws to fell trees and even into fully portable tools such as handsaws or hand drills. In fact, the very reasons which originally compelled us to adopt the factory system as such are no longer cogent, and we are in some matters on the point of returning to cottage industry.

If indeed the logical place for the invention and use of the electric motor had occurred fifty to one hundred years earlier, it is doubtful whether the cottage system of manufacturing, with both its advantages and disadvantages, would ever have died as completely as it did. It might now be resurrected in certain limited fields.

Very possibly the countries which are only now coming to a highly industrialized stage may experience much less of the trend of population from the village to the industrial city than England and America did in their time, and it is quite possible that this may soften for them some phases of the fell impact of modern industrialization.

The introduction of the fractional horsepower electric motor, even though it is an inverse invention, is indeed an idea which would naturally come from the side of the entrepreneur, and thus might have turned out to be suitable for a mass organization of invention from above. It is hard to see how this could have happened, however, with the series of inverse inventions which constitutes much of modern electronics. That the vacuum tube, which made its first practical appearance as a device for wireless transmission and reception, should prove to lead to the computing machine and the industry of automatic control is no simple guess which could have been made from the outside, but demanded a repeated scientific and engineering insight into the nature of the device.

Once this insight had been obtained, the various detailed problems to which it has given rise became subject to a mass attack, and this attack continues. However, the original idea that the vacuum tube might fulfill these functions is the result of just that sort of purposeful day-dreaming which is not very much encouraged in the closely-knit and highly organized type of mass research. It is precisely because certain engineering laboratories have not ridden the wave of the future to its ultimate breaking on the rocks that these innovations have come to pass.

The vacuum tube is an invention of far greater depth than its first users realized. As I have said, its first function was to assist the sending and reception of wireless mechanisms. It has only been seen in the last twenty years that the use of the vacuum tube for purposes of amplification has divorced from each other two engineering problems which had always been associated inseparably, and has made possible their remarriage on a new and sounder basis.

These problems are that of the transfer of energy and that of the transfer of signals and information. With the aid of the vacuum tube, an arbitrarily weak signal, provided only that it stands out clearly from its background of noise (that is, of signals which carry no desired information), can be used to regulate the performance of an apparatus working at any desired power level.

For example, the classical microscope brings to the eye only a portion of the light with which the observed object is illuminated, but the new flying-spot television microscope is able to intensify the illumination of the image to any desired degree. The feedback process on which the automatization of industry depends is only practicable because extremely faint signals engendered by the unsatisfactory performance of a machine may be built up to a point where they initiate an effective process for the correction of the performance of the machine.

When we see what is implied by the separation of effective signal from effective power, this is clear and almost trivial, but it was not a trivial step to see that all this was really implicit in the properties of the vacuum tube itself. It is easy to stand the egg of Columbus on one end, after Columbus has shown the proper way to do it. This takes one brain, not a thousand cramped half-brains.

The great difficulty with highly organized research, with its preassignment of the functions of the individual scientist, is that it is a bulky and lumbering machine, which cannot easily be redirected to meet the shifting needs of the world of ideas. The great organization, which has accepted a purpose involving the expenditure of millions of dollars—megabuck science—is not easily turned aside; nor are the best administrators of large projects likely to be the most fertile originators of new ideas. Thus megabuck science is not likely to be a good midwife or nurse of new ideas.

New ideas are conceived in the intellects of individual scientists, and they are particularly likely to originate where there are many well-trained intellects, and above all where intellect is valued. You will not find much invention in a poverty-stricken coolie population. On the other hand, if there are enough sneers at eggheads, many intellectual eggs will fail to hatch, or even to be laid. Our McCarthys and McCarrans cannot continue indefinitely to storm at the intellectuals without making many a person wonder, Is the certainty of this obloquy worth all the risk? It is true that the best and the most devoted scientists are probably moved by interests and curiosities which are so firmly established that it is difficult to tear them away from their natural direction of effort, but even though this is the case, the general statistical effect of an anti-intellectual policy would be to encourage the existence of fewer intellectuals and fewer ideas.

It is not enough, though, to have ideas in the minds of individuals if there is not an adequate means for communication between these minds. On the one hand, journals and, in past times, academies have enabled men of similar thought to come together and to multiply their ideas by cross-fertilization. On the other hand, a policy of secrecy or even, what amounts to almost the same thing, a policy of the minute subdivision of intellectual activity does not give an innovation a fair chance to be born. How much has mankind lost, I wonder, by the language of concealment in which the notebooks of Leonardo da Vinci were written, and the actual mislaying of these notebooks for centuries, until well after the time at which most of his ideas had been rediscovered?

Even if we grant that some secrecy can be obtained for inventions, the question of the precise degree of secrecy that can be obtained in invention is an extremely difficult one to answer. It is, however, an extremely timely one at present. Let us take the atomic bomb as an example. Here I am speaking as a member of the general scientific public, without access to specific information concerning the atomic bomb, and with less than no desire to have access to it. The early history of the ideas leading to the atomic bomb goes back through the work of the Curies at the beginning of the century, and the earlier notions of Becquerel on the one hand; and, on the other, to Einstein's identification of matter with energy on the basis of a highly abstract dimensional reasoning.

By the time the century was well along, the following things were known: that radium and similar metals seemed to undergo a sequence of transmutations from one elementary form to another on a purely random basis; that in this series of transmutations, radiation was emitted, of at least three types; that matter and energy were quantities of essentially the same nature, but connected by a transformation factory so extremely great that if matter were to be made to dissipate itself entirely in energy, the amount of matter in a desk-weight would be more than enough to propel an ocean liner across the Atlantic.

With these ideas already accepted, the ground was laid for further investigations concerning the transmutation of one kind of matter into another, and concerning the transmutation of matter into energy. The scientists could not be expected to be content for a long period with a transmutation that was merely fortuitous and beyond control. Gradually it became apparent to the scientific world that the atoms of the elements were made of two parts: a rather light outside mist of rotating electrons, and a central nucleus of vastly greater mass and consequently of vastly greater energy. It also became clear that the known radioactive phenomena concerned this nucleus rather than the peripheral electrons. As to the peripheral electrons, they were the locus of the ordinary chemical properties of the atom, but the same cloud of electrons might be, and in fact was, associated with more than one nuclear configuration.

In fact, it became clear that the atomic weights of what were then known as elements were not absolutely fixed, and that each species of element might consist of a number of different isotopes, to all intents and purposes identical in chemical properties, but different in atomic weight because of a different nuclear structure. When this phenomenon of isotopy was discovered, it also became clear that isotopes, for all their chemical equivalence, might have widely varying radioactive properties.

The first experiments leading most immediately to the atomic bomb concerned an artificial transmutation of the elements in which we were no longer content to wait for a fortuitous event to produce the change but tried to cause it ourselves. To do anything as powerful as this in accord with our ideas, both then and now, meant to subject the nucleus to extraordinarily large concentrations of energy or, what is equivalent, very high temperature.

The earliest successful artificial transmutation of the elements was carried out in 1921 by Cockroft and Walton in the Cavendish Laboratory in Cambridge, England. It was a moderately large-scale experiment, but nothing like the million dollar experiments of which our present physicists talk so complacently. I remember seeing the apparatus. The large vacuum tube in which the experiment took place was far from being an elaborate triumph of the glassblower. It was made of cylinders of glass not more than one or two feet high. These cylinders were piled upon one another in such a way that between each consecutive pair there was a flat piece of glass of the size of a large window pane, with a circular hole cut out of it by, I imagine, the use of an ordinary glasscutter. These various parts were held together and rendered vacuum-tight, not by the use of the torch of the skillful glassblower, but by the employment of a certain glorified black sealing wax of remarkably low vapor pressure, known as Dekautinsky's cement. I am not in a position to judge what the cost of the more interior and essential parts of the apparatus was, but it did not seem to be anything that would seriously strain the resources of even a small American college.

None of this work was secret, nor was there an obvious reason why it should be secret. However, it had already become obvious that the energetic problems of controlled nuclear changes would have to be considered, all the more especially because the work of Einstein on the relation between mass and energy had made it clear that, for good or evil, we were on the threshold of a new and potent source of energy. When the investigations of Lise Meitner, just before World War II, had given a clear suggestion of the way in which these new possibilities might be exploited, the fat was already in the fire, and no further fundamental atomic secrecy was possible. The best secrecy which could still be expected was the limited, dated secrecy of armies, concerning the specific use of means and those details of construction of a relatively ephemeral nature.

One of the main points of any policy concerning the limited secrecy which is really significant in war or in competitive business is that it must have a definite purpose. A second, almost correlative point is that such a policy must be really capable of being carried out effectively, and that it should not waste effort attempting to do what one might expect in advance to be impossible.

As to the first, the purpose of any policy concerning secrecy and information should be to establish some sort of an advantage in conflict. In this, it is important to consider not only what secrecy does to the effectiveness of the other side, but also what it does to that of our own. Any damming up of the freedom of flow of information is bound, sooner or later, to have a disadvantageous effect on us ourselves in the matter of information. This is particularly the case because, in a period of invention and creative work, the precise channels by which information may become effective in invention, discovery, and the new use of our will cannot be given in advance.

To channel information and to classify it is without doubt a necessity in times of great emergency, in which it is more important to make effective use of information already at hand than to acquire new information which may be serviceable at ten years' range. On this basis, it is also possible to defend the diversion of fundamental creative effort into the more immediate but less creative channels of highly organized research. These measures belong to an emergency strictly localized in time, however, and not to a continuing emergency. They belong to the sprint and not to the marathon race.

In a sprint, we do not expect that a man will have to keep up a physical balance between the oxygen inhaled and the oxygen used in his tissues. In the time allowed, he cannot possibly use up his reserves, and he can therefore go all out. This is not the case in the long-distance race. There the runner, notwithstanding that he is not keeping up a true equilibrium in food and water, or even in oxygen, must approach to a method of using his breathing which will be continually effective over the whole race. Within a few hundred yards his reserves will have largely vanished.

Perhaps the brief emergencies of the closed wars of the past could be survived on the basis of a moratorium on fundamental research work and a quick activation of our powers of making past research work available in new technical improvements. This moratorium ceases to be possible, and certainly to be advantageous, in a struggle lasting decades or centuries. It is necessary for us to watch very carefully lest the methods we use to keep a concert pitch for immediate activity do not, in the long run, exhaust the creative reserves on which we must depend year after year. In fact, we may say that, humanly speaking, if we have so organized our flow of information that there are no minor leaks, we have probably channeled it so narrowly that the healthy internal growth of our own technique is hampered, and that we had better reconsider our action for the long haul.

So much for the problem of balancing secrecy and the internal integrity of information. The detail of this balance is difficult to discuss, because the very material on which an effective discussion could depend has been closed and blurred by the policy of secrecy itself.

Let us discuss in some detail, not the desirability of secrecy to prevent a possible enemy from acquiring advantages, but the general possibility of this secrecy and the degree to which we can act in the hope that such secrecy is nearly complete. I have already said that by the time an invention reaches the stage in which it is commercially advantageous to press its immediate application, the fundamental basis of the invention is very widely known.

What I have said about industrial applications applies equally to military applications. Quite apart from the leaks of our own inventions that occur from our own side, we must expect a certain moderately high rate of rediscovery of our own tools and techniques by our antagonists, actual or possible.

Unquestionably, this parallel growth, while it is extremely important, has not been enough to account for the full development of atomic techniques by the Russians, for example. We have chapter and verse of certain acts of espionage which have seriously cut down the time in which they might have caught up with us by their unaided efforts. Nevertheless, as far as the man in the street goes, and,

I suspect, as far as the Pentagon goes as well, a large part of the head start which we have supposed our inventions have given us must be discounted.

Much that we have invented was quite ripe for invention anywhere in the world. Perhaps if we had kept our secrets perfectly, we might have parlayed a five-year lead into a ten-year lead; perhaps, though I doubt it very much, into a twenty-year lead; but nothing could be more certain than the fact that we could not have parlayed it into a lead of half a century.

Now, at least in part by virtue of obtaining our secrets, the Russians have cut our lead to not much more than five years. We talk, for the benefit of the man in the street, of the theft of the atomic bomb. The important thing to realize is that the word "theft" brings up moral issues from a one-sided point of view, and from a point of view in which we could not expect anyone else to concur who was not bound a priori to our interests.

If Russia had had the head start on us in atomic weapons that we had on them, we should have made every effort to plant our own Fuchses and Rosenbergs in their laboratories, and if our administrators had not done so, we should have considered them very remiss indeed. Until international affairs are on all sides a matter of cooperation instead of conflict—and this day, for all its desirability, is very far from here—no nation can sit passively by while another nation develops a weapon which will destroy its independence of action.

This fact we must consider with open eyes, and although it is perfectly reasonable to hang a spy when he is caught, it is quite unreasonable to expect that there will be no spies, or that the other side—any other side—will refrain from trying to make use of them. Some of these people may be found among the adventurers that every war or struggle casts up on the beach of history and whom every side uses, disavows, and ultimately destroys.

There will be others who, from an enthusiasm which may be national patriotism in the narrow sense, and which may be devotion to what they conceive to be a principle above nationality, take the extreme risk and even constitute themselves the kamikazes of espionage. We may deplore this, and we certainly must fight fire with fire, but it is unrealistic to waste our moral righteousness in a general way against an enemy which takes such action, just as it is to waste moral indignation against a kamikaze fighter or to expect that our feeling against kamikazes will stop an enemy from using them, and that it will render unnecessary the protection which we accord to our airplane carriers by strong antiaircraft batteries and by fighter planes.

Some secrets will get into the hands of the enemy, and a policy which altogether ignores this expectation is not realistic and not conducive to our own safety. We must expect of secrecy only what secrecy can reasonably perform. Even more than this, as far as this secrecy itself goes, we must determine its degree on the basis of an objective consideration of both its advantages and its disadvantages.

I do not in fact believe that we have made an adequate objective examination of the possibilities and proper balance of secrecy, and I know that even if this realistic balance may exist somewhere in the Pentagon, it does not exist in the minds of the men in the street or the newspaper editors at large.

There is a general present tendency to suppose that invention exists primarily for the sake of the individual profits made by exploiting such inventions, or for their competitive use by different groups and even by whole countries as a source of power and control. Invention is conceived as a means by which the wills of others may be made subservient to one's own. If this were the only use of invention, or even the main use of invention, the problem of secrecy, complicated as it would still be, would become relatively simple. What complicates the problem is that invention also fulfills a fundamental noncompetitive human need, and that the same invention will fulfill different needs at different times.

At the period in which new intellectual ideas are fighting themselves out in the human mind, it is often quite possible to foresee what can be made of this new discovery as an individual possession or as a means for asserting national dominance. Many inventions and discoveries have it in them to be as imposing as an elephant and as dangerous as a lion, but neither a lion nor an elephant is a particularly impressive lump of flesh in utero. Yet it is precisely this uterine period during which an invention, like a fetus, might be subject to some ideal control and is, in fact, a rare and highly improbable accident. For it is one of the paradoxes of probability theory that what happens is, in fact, always enormously improbable in its full concreteness.

That you or I should exist means that a certain ovum among five hundred that actually leave the ovaries, and among a vastly larger number that are formed in the ovaries and cells, should meet and be fertilized by a single spermatozoan out of many millions. No other ovum and no other spermatozoan would in all probability have formed precisely that combination of genetic traits needed for the formation of the particular individual. Yet when the individual is actually present with all these traits of nature and of nurture, he is a highly real and important fact quite conceivably likely to leave a mark on the world for a very considerable time.

Similarly, the events leading up to the conception and the parturition of a new idea may be so extremely unlikely and unpredictable that a trivial accident could have prevented this combination from arising at all. At any rate, it is likely to arise at a particular time and in the mind of a particular individual in a manner far too fortuitous to be the subject of a reasonable gamble. Our employers do not invest in embryos, and our capitalists are no more able to invest in embryonic ideas.

Thus, precisely at the time that an idea might be controlled, if only by a process of abortion, it is of no predictable interest to anybody to control it. Later on, when the idea has reached a viable childhood or even a lusty youth, it is no trivial task to make it as if it had not existed. By this time, nothing less than a spiritual murder will touch it, and, like other murders, spiritual murders will out.

The point is that by the time an idea is ripe for any predictable economic or military use, a lot of people will have seen it and it cannot likely be eliminated. This means that ideas that may serve as a reasonable basis of an economic investment must have already been recognized by many traces of their existence that are widely scattered over the community. By this time, the phenomenon of multiple invention is not merely an outside possibility, but a very definite probability.

Furthermore, the secret of the existence of an invention at the late stage at which it becomes inviting in an economic or a military way is already a very shallow secret. The ideas are there, and although it still may take a certain amount of acumen to discover their aptness for a special purpose, it is scarcely likely that many decades or even many years will pass without persons of that degree of acumen arising repeatedly over the whole range of society which has been subject to even a remote contact with the new ideas.

Unfortunately, the bonds which have linked scientist to scientist and mind to mind in different countries have received a certain loosening of late. The Iron Curtain is upon us and is supposed to represent a two-way barrier. Yet it is a barrier which remains quite permeable to books and to periodicals, and the state of knowledge, let us say, of physics, on one side of this barrier is not very vitally different from that on the other side.

Short of an extremely radical restriction of publication and of scientific contacts, which will not merely cover those military and economic tools which are actually usable at present but will extend to all nascent ideas whatever their nominal field of applicability, the Curtain will remain a very permeable membrane. The cost of making this Curtain impermeable to all ideas, whatever their field, and whether they are nascent or mature, involves the secondary cost, which is far from trivial, of slowing up or abolishing intellectual development in each moiety of the world separately. Ultimately, it involves the abolition or the extremely heavy restriction of all scientific progress whatever. Short of this, secrecy is a snare and a delusion.

It is a commonplace of history that our good generals always fight the last war, and the generals of average ability, the war before the last. What is true of weapons of steel and explosives is equally true of weapons of ideas and of policy. The continued existence of our country or even of humanity depends on the sudden and thorough acquisition on the part of our policymakers in war, in diplomacy, and in every sector of our life of some consciousness of the plane of ideas.

It is not enough to talk secrecy and security, or even to demand them. We must have some adequate ideas of what can be attained and what we wish to attain, and some well-thought-out policy of attaining these goals. This idea and this policy are still in the future. What belongs to the immediate present is the blight which secrecy, along with unhuman thinking and with rigid, overgrown organization, have imposed on our men of ideas.

In this age of Ragnarök, in this epoch when winter after winter of the human spirit follows without an intervening summer, the true creative scientist, the originator of ideas, must remember to honor his mission in life. For there is at present a fundamental opposition between the spirit of the free creative scientist, the originator of ideas, and that of the scientist who is working in an organization which is primarily adapted either for the commercial exploitation of ideas already introduced, or for those later stages of the act of creation and discovery which can be reduced sufficiently to rote to be treated as a commercial enterprise. Naturally, the two sorts of scientists themselves may perhaps live peaceably together, but it is hard to see how their values can correspond. It seems to me that their coexistence can lapse from time to time into the covert opposition behind it.

To the scientist outside the great organization, the glorification of the great institution by those within it sounds like the attempt by the fox who has lost its tail in a trap to establish a new fashion of bobbed tails. One of the reproaches most commonly made against the independent scientist, both by those foxes who have already sacrificed their tails and by the would-be employers of science, is that the independent and individual scientist is undisciplined. Undisciplined he may be if undisciplined is taken merely as freedom from the shadow of external restriction and of possible external punishment. On the other hand, he must be a man of profound discipline if he is able to seize the vague and formless hints of ideas which are all he has to work with, and to reduce them to cogent and to manageable form. One does not invade the realm of chaos and reduce it to order without a very compelling internal need for order, nor does one accept with impunity ideas which are self-contradictory and incapable of being brought into an orderly form. Thus it is a part of the intellectual discipline of the creative scientist, when he finds that his hunches are really empty and that the hope of order which he has been chasing is nothing but a will-o'-the-wisp, to put this will-o'-the-wisp aside and leave the bog for firm ground under his feet.

This part of discipline is both real and very obvious, but it is by no means the whole discipline necessary for the scientist. As the exact counterpart of that solidity of outlook which leads him to turn aside from the untenable, once its untenability has been proved, there must be the other and less popular discipline which leads him to continue to entertain an idea until its formlessness or incorrectness has been thoroughly established.

The history of the mathematical work of the Jesuit, Geronimo Saccheri, shows the need of not rejecting paradoxical ideas too soon. In the eighteenth century that speculative thought which had taken the postulates of Euclid and the Euclidean derivation of geometry to be final for more than two thousand years began to come to grips with its problems once more. As an axiom of mathematics was supposed to be a truth which one could not doubt without contradiction, there was a renewed attempt to prove the axioms of

Euclidean geometry to be of this nature. Of these axioms, the stubbornest and hardest to establish as irreplaceable was the Parallel Postulate, which asserts that in a plane and through a point lying in this plane, but not on a given line on the plane, one and only one line can be drawn not intersecting the given line.

If we hope to show this to be a true axiom, we must show a valid reason for discarding each of two alternative postulates, one of which allows several lines to be drawn through the point, not intersecting the given line, while the other allows no such line to be drawn. In order to show that the acceptance of either of these alternatives may lead the ship of geometry onto the rocks, Saccheri began to draw conclusion after conclusion from each of these two hypotheses. The results that he obtained were most grotesque, but although he regarded them as unacceptable, they did not lead in either case to anything like a manifest contradiction.

Later mathematicians such as János Bólyai and Nikolai Lobachevski followed Saccheri's line of thought to a different conclusion. In fact, they were able to show that each of the two new postulates led, not to a contradiction, but to a geometry different from that of Euclid, and in fact to one of the geometries that has since been called non-Euclidean. To have discarded these non-Euclidean hypotheses merely because of their strangeness, without any successful method of carrying the one or the other to a contradicting conclusion, would have been a sign of intellectual incompetence if not of intellectual cowardice. Thus it would have been undisciplined to abandon the study of the new geometries simply because each one of them demanded a different adjustment of our imaginations than that which we had been in the habit of making.

In my own work I have had to face a similar quasi-moral issue, not once but many times. For example, in the geometry of the Brownian Motion, I was forced to entertain in a physical theory the notion of continuous nowhere-differentiable curves. These curves had long been known to have a mathematical existence, but they were regarded as museum pieces entirely outside that physically significant part of mathematics which was thought fit to be applied. In order to arrive anywhere with my ideas, I was forced to entertain the concept of such curves in a truly physical situation. To have refused to do this would not only have excluded me from a main part of my later scientific results, but would quite rightly have been considered as incurring a loss of my intellectual stature. It takes discipline, not merely to abandon that work which must be rejected, but not to abandon that which still merits consideration.

Thus I take my stand in the support of innovation. Some readers may wrongly conceive that I maintain a general notion of progress in science. That science is getting bigger and better in every way is not one of my tenets. Nevertheless, in so far as science has printed and written records, and in so far as the bulk of these records gets larger and larger from year to year, the field which must be searched by the scientist is continually increasing, and this whether he wants it or no. Science is an irreversible process, short of book-burning or other means of reducing the stature of the scientist. We cannot with impunity act as if the bulk of literature and thought which represents the achievements and the speculations of past ages were not there. For better or for worse, we start where we are, and the good old times are as dead as Marley's ghost.

This is a situation which does not essentially differ in science and in the arts, though there are certainly those who wish to turn back the clock. Let us suppose that with the sweep of his wand a wizard could bring Newton down among a crowd of scientists of the present day. By this I do not mean the copybook Newton, who has existed chiefly for the purpose of having his name inscribed on buildings, but the real Newton of early middle age, at the time at which he did his best work. Of course, he would not be at home instantaneously in the researches being done at the present day. The first things that I can imagine him doing would be to get hold of the books of Gibbs, of Einstein, and of Heisenberg and to study these carefully. It might well be that he would find flaws in some or all of these, and that he would reject them.

What is certain, whether he accepted or rejected them, is that he would not ignore them, and that some of the work which he would do would throw light on the moderns. For a man who had the frankness and courage to say, "If I have seen further [than others] it is by standing on the shoulders of Giants," would not hesitate for a moment to climb even higher on the shoulders of our latter-day giants.

If Euripides could be brought down to earth with all his creative powers and his consciousness of guilt and sin, it would only be a matter of days before he would realize that the later-born had been investigating these problems, and to some good effect. It would not be long before he would become aware that Freud had existed, and he would read the works of Freud with the greatest interest, whether he might accept or discard them. In any case, a Euripidean play written more than a year after his incarnation might indeed avoid the Freudian jargon, but it could not possibly steer clear of the Freudian repertory of ideas.

If Leonardo were to be among us and were to visit the museums and see how the intermediate ages had bid fair to exhaust most of the possibilities of purely pictorial quasi-photographic art, I suspect that he would become as bored as other great artists have done with those valid traditions which have lived so long that they have become clichés. I do not presume to say exactly how he would experiment with new techniques of painting and of statement, but I am morally sure that if he stuck to his last we should find him exploiting some traditionally quite unorthodox methods, and even as the patron saint of some modernistic school.

This leads us to a very interesting artistic question, the question of the art of the historian, the translator, the antiquary, and the writer of pastiches. The translator of Dante must talk as Dante did, and must use no idea which would have been foreign to Dante. However, by this very fact he cannot be a Dante, for Dante would not have scorned any new idea that might have come to him, even though, in fact, he only used ideas stemming from the ideas around him and combined with his own variations of them. At any rate, he was under no compulsion to confine his ideas to the range of those of some existing Dante. Ralph Adams Cram felt himself under a necessity to design modern buildings with the same repertory of methods and religious conceptions as those of the architect of Chartres Cathedral. In doing this, Mr. Cram was working under conditions which would never have been accepted by the architect of Chartres Cathedral. The architect of Chartres would not have scorned new building methods or striking new forms which might have come to his attention. He was medieval, but he was not a medievalist. It is in accepting a dead tradition as something closed and finished that the medievalist in latter-day architecture shows his Cramnation.

Noncalculable Risks and the Economic Environment of Invention

9

A phrase much in favor among the policymaking circles of the present day is "calculated risk." It has a valid meaning and is, for example, the proper basis of the insurance industry. When we consider a single individual, he may die tomorrow or fifty years from now, and we cannot be too sure which.

However, if we take large numbers of individuals, they can be divided into various classes by a medical examination, a study of the longevity of their ancestors, a classification of the work they do, and so on, and for each of these classes there are careful statistics which, covering large numbers of cases, will give a pretty accurate estimate of the number of members of the class whom we may expect to see alive after a given time.

This work of examination and statistical classification represents the interlocking task of the insurance physician and the actuary, and it furnishes an excellent basis on which to bet, so that a life insurance company with a wisely spread clientele can be quite reasonably sure that it will not be put out of business by any single fluctuation of its data or any single catastrophe short of a major outbreak of the Black Death.

Now we hear news statements to the effect that when the United States used the atomic bomb at Hiroshima, it took a calculated risk. Who, may I ask, were the actuaries who determined this risk? To employ the atomic bomb involved an estimate not merely of its killing power, but of its emotional impact on the Japanese and, even more, of its emotional impact on all those members of non-European races who were quite sure that the United States would employ it differently against Asiatics and blacks.

Moreover, the use of the atomic bomb involved the contemplation of the position of delayed reprisals against a country which had already sanctioned it. The technique that might lead to the manufacture of this bomb or of even more powerful weapons was still in its infancy, as indeed has been shown by the later use of the hydrogen bomb. Our secrecy regulations, in order to be adequate to keep the bomb from our enemies, had to be far tighter and more enduring than any such secrecy policy had been in the past, and had to stand up against the imperative demand of our possible antagonists for freedom to determine their own national policy in the future. Where, I ask, was the actuarial material to be found for judging all this? The atomic bomb in its military, social, and political implications is a good example of the fact that a really new discovery must always invoke important elements of incalculable risk.

Actuarial work can only be done when we have a sufficient number of instances, so that the degree of deviation can be assessed. This is never possible at the beginning of a radically new course of conduct. The result is that there is neither a calculated risk nor a calculable risk, and no amount of mathematical acuity can make up for a small statistical knowledge. "Calculated risk" has often become a cheap catch-phrase, and is very commonly meant to fool the public.

Some years ago, an extremely interesting example of the question of calculable risk came to my attention and the attention of Professor John von Neumann. I had for some years been working over the mathematical theory of prediction, and I had devised methods which could be used on numerical series such as stock-market data. I was loath to exploit this in a quasi-tipster service, for I knew the blind confidence with which many clients of such a service accept the competence of the statistician, and I knew that at that very time when statistical guidance was more to be sought, some of the fundamental dynamic factors in economics were undergoing a rapid change. Therefore, I could not look upon the proposed commercialization of my own ideas with a clear conscience.

At this time, a scion of a great industrial family visited me in my office. He was eager for me to go ahead with the problem of stock-market prediction. His notion was that even a one percent improvement in stock-market prediction would regulate the investment of so many million dollars that it would become a major factor in economic life.

When we brought this to the attention of von Neumann, he pronounced an opinion with which I concurred completely. For an improvement in stock-market prediction to be of any commercial value at all, it would not be enough for it to be serviceable in some uncheckable way to the tune of millions of dollars. Those who employ it must have some way to be aware of its proved serviceability.

The point is that there are many factors in stock-market prediction of which we are not fully aware, and these are likely to make changes running up to several percent in the merit of the prediction. Thus an apparent one percent improvement could be merely due to chance secondary economic factors and would not be demonstrably due to the merit of the method of prediction employed. Even if a real improvement in prediction were there, I do not see how we could know it in any reasonable time and without an unreasonable expenditure in statistical investigation.

For us to be able to use a statistical estimation method to cut down a calculable risk, it is necessary for the improvement produced by this method to be clearly calculable and even directly visible. In other words, any effective process of regulating one's investments must be a controllable process. Like all such, it gains enormously if it contains a feedback by which its past efficacy may be judged in planning its future. Anything short of this is not a satisfactory basis for policy, and if we count on a policy based on so slender a foundation, we are extremely likely to ruin ourselves.

This consideration is of the first importance when we take up the question of calculated risk in invention. The most critical stage of invention, as we have said, is the change in intellectual climate which produces and is produced by a new idea. This may be of untold value to the community, but in the essence of things it is not subject to actuarial work.

It is only later, when the idea is spawning all over the community and one man after another is informing himself of its potentialities, that even the purely intellectual experiment of a highly theoretical actuary becomes possible. At the earlier stage there may well be more to gain by following the new lead, but our uncertainty as to whether there is more to gain or not is so great that it represents an abominable investment. The limited industrialist must establish for himself not simply that an idea will pay off to somebody, somewhere, but that it will pay off to him within such a time that he can reasonably count on a payoff in auditing his books.

Thus the industrialist must keep away from the first, the seminal stage of inventions and discoveries, and must wait until the later stage of invention, when what is not done by one man will almost certainly be done by another. Here he stands to win less by a great coup, but he will have a much better estimate of whether he is winning and what he is winning. He has hedged his bets; and if he has cut his remote and inestimable possibility of fabulous gains, he has reduced a wild speculation to a controllable business.

One of the great difficulties of policymaking in any field is that policies cannot be made indefinitely by dead reckoning. To sail a ship by dead reckoning alone, without sight of the sun and stars, without the use of the lead, and without the possibility of a distant view of the coastal landfall means ultimately to run up on the rocks. The sextant, the radio-compass, and the lead are all of them feedback apparatus, by which we can check a predicted position against an observed position, or at least against some indication of an observed position. Similarly, a business or government department cannot form or follow a policy effectively without some repeated index of the effectiveness of this policy. All this seems reasonably simple, and must form the guide for our short-time and even middle-period policy.

The difficulty with this method of proceeding is that its effectiveness decreases with the length of time it takes for a policy to make its consequences known. In administering a redwood forest, which may well take more than a thousand years for a crop, we shall get nowhere if we do not consider at least the possibility of intervening forest fires, or changes of national ownership, and even of a future human race which in some ways develops some very different demands than those of the present time.

The problem of predicting infrequent contingencies long ahead is a mathematically difficult one, and although Emil J. Gumbel has done something with it, I do not think it is a problem for which we have an adequate answer. In New England, the biggest risk to be stood by a dam in the course of a year is almost certainly a statistically estimable excess of the rainfall over the normal. In a century, it is likely to be a West Indian hurricane, or perhaps an earthquake. Over a period of a thousand years, we simply do not know. The long-time contingencies belong to a different population than the short-time contingencies, and an extension to a long time of feedback measures based on short-term experience may fail to work. However, with better and better knowledge of history and geology, with an examination of the varves, the mud-layers thrown down in ancient lakes, and similar new undertakings, we may ultimately learn to estimate those long-term strategies which may be best in the construction of the dam.

To sum up, every continuing undertaking must be regulated as to its performance by its results. These results constitute, from the information point of view, something which is strictly analogous to what is called "feedback" in a control machine. The theory of economic and political behavior which is now in vogue in the United States, and which has received a quasi-official sanction, consists fundamentally in a very definite view as to the nature of a feedback which is regarded as sufficient for all social phenomena, and which is enjoined upon us.

This feedback is the economic feedback of private enterprise. The counters by which the success or failure of an undertaking is to be judged are either money or something translatable into money, and the impingement of the gain or loss of this economic quantity upon individuals or corporations is supposed to be enough to regulate all undertakings. That is, when the returns are called in and we know what profit or loss an undertaking has incurred by the commercial exchange of buying and selling, by the interest on capital, by the expenditures necessary for replacement, maintenance, and depreciation, we are supposed to have a complete guide to the degree of success or failure of the undertaking.

If we look on the returns of business in the light of history, we shall see that this bookkeeping, this monitoring of success, is a relatively short monitoring. The present generation has no illusions as to the absolute permanence of monetary values. We have seen a great depreciation of currency and, indeed, an inflation covering much besides currency in the strictest sense. What we are not always conscious of is that this devaluation, or what amounts to the same thing, by the loss by wars, famines, and the like, is a presupposition of the capitalist system itself.

Suppose that we are to consider a Roman coin of the time of Christ having a purchasing power of $1.00. Suppose we consider it put out to compound interest at the very modest rate of 2%. What would it amount to now? The order of magnitude of the total savings would be something like a million billion dollars. Manifestly this is nonsense, and manifestly the answer is that no banks have been running under stable conditions for any large fraction of two thousand years. Most of the interest which we receive is conditioned by the repeated destruction of goods and fortunes that has taken place in the course of time. Any really stable social system without intervening catastrophes would offer a rate of interest so low that no individual would be much concerned with investments over periods comparable with his own lifetime. That is, the rate of interest of business and the capitalistic system are conditioned by the fact that our businesses are relatively short-time undertakings, and are not able by their very nature to pay much attention to the long-time secular interests of the human race.

In the construction of control machinery, we often, and even always, find that the continuing function of our machines is dependent on a number of feedbacks on very different time scales. For the purpose of argument, we shall limit ourselves to the case of two feedbacks, one of which is of very short duration over the past, and one of which is of vastly longer duration. We shall call the first the immediate feedback, and the second, the secular feedback. Let me give a simple engineering example of a system with both immediate and secular feedbacks.

The bringing down of an airplane by the fire of an antiaircraft gun is essentially a statistical problem. We cannot hit the plane every time, but we want to hit it as often as possible, or at least so that the damage done will be as great as possible. In order to do this, we take the observation of the plane over a relatively short past time, and by certain manipulations of our previous observation we determine the best direction in which to shoot so as to bring the shell to where the plane will be at some time or other in the immediate future. The direct data which we use will be distributed over the past motion of the plane for seconds or fractions of a second. However, these are not the only data which we use.

In order to determine the policy by which our gun shoots, we must know something of the statistics of plane motion. An adequate knowledge of such statistics will require extensive records of the motion of the type of plane against which we are firing while it was engaged in the type of activity in which it apparently is now engaged. The very least material that will be of much value to us would involve a record of the plane, or similar planes, for minutes or hours, but instead of this we would probably use a synopsis of records of flight of enemy planes over days or weeks or months. It would be possible to give a computational process (which could be mechanized) which would allow us to take these data, to examine them statistically, and to work up the policy to be used in the short-time feedback fire control problem. This feedback process is of a secular nature.

When we come to human policies, for which our short-time accounting is annual, or perhaps over five- or ten-year periods, our long-time accounting can only be the accounting of history. It depends, as does the accounting in the case of aircraft fire, on memory and on records, but here the memory and records are long-time memories and records of the human race. Obviously, the bookkeeping of a particular firm or a particular individual is too short and too narrow to take these factors into account. How, then, are we to account for them?

Notwithstanding the apparent claims which some of the extreme believers in the capitalist system make, that rewards and punishments of money and its lack are able to deal with all problems of administration, business does not really believe this itself.

A very important type of business is the insurance business, in which, for an appropriate sum of money, the insurer undertakes to assume and make good certain risks and losses of the insured. It would not, in fact, be going very far out of the way to say that all the conduct of business, as far as policymaking goes, amounts to an insurance within an organization of the risks inherent in the conduct of the organization.

Now, as I have mentioned before, while insurance companies will take care of a very great variety of risks for which actuarial tables are available—that is, of calculated or calculable risks—there are certain types of risks which they are not willing to touch at all. If you read the small print of an insurance policy (or of certain other contracts such as a steamboat ticket or an airplane ticket), you will find that the coverage makes an exception of "Acts of God and of the State's enemies." Leaving out the special risks of war and the like, these cover catastrophes of so rare and so total a nature that there are no adequate actuarial data to cover them.

We have recently had a not-too-great catastrophe of this character in the form of the Worcester (Massachusetts) tornado. In fact, serious as this was, it was not large enough to put any important part of the ordinary facilities of insurance and the like out of action. Nevertheless, the necessary succor to the injured and the bereaved was not the primary work of an insurance company or of any business organization, but of the Red Cross.

The Red Cross and other such organizations for the care of victims of emergencies appeal to sanctions outside of the capitalist system. They appeal to religion and the church, and to our general sentiments of humanity. They appeal, in other words, to institutions which can be, and generally are, of long duration compared with any individual business. The capitalistic system of sanctions and of feedback which we actually support is one which is deeply modified and mitigated by the concurrent existence of charitable and religious institutions and other social institutions where the profit motive is not the main motive.

If it is impossible for the short-time returns of business to take care of rare and unpredictable catastrophes, it is equally impossible for it to take care of rare and unusual events accruing to the benefit of humanity. I have said that religion and the church deal with long-time risks and policies, whether or not it can be strictly said that they deal with the eternal. It is therefore appropriate to use the language of religion in speaking of long-time great and unexpected benefits, and these the church calls Acts of Grace.

A great new intellectual discovery, although it is not the only sort of event which can be called an Act of Grace, certainly belongs in this category.

The remarks which I have already made about feedback indicate that the protection against unusual catastrophes of the Act of God variety depends upon those long-time secular policies which cannot receive a visible check of their validity within so short a time as a human life or the expected duration of a common undertaking. It is therefore necessary to study history in order to know how to provide for these contingencies and to see what policy in past history would have led to the best conservation of human and humane interests under similar circumstances, and whether present conditions are such that this judgment can be expected to continue to be valid.

The acceptance of such secondary feedback as, for example, the necessity of charitable relief after a tornado or a flood because we see that, according to history, such action has on the whole led to favorable results, and because we accept the present example in the light of this history, is what one terms religiously an Act of Faith.

The thesis that I am maintaining and have maintained in this book is that the continued existence of an atmosphere in which fundamental science can develop to such an extent that it can fulfill not only our hopes but our needs, depends on the faith of the community that the work of the intellect is worth promoting and that institutions devoted to the furtherance of the atmosphere in which this work can take place represent a public interest.

It also involves what I have already mentioned in connection with the devotion of the scientist, namely that he himself must be a dedicated man. The environment in which he can flourish must be left to institutions which are essentially built for the ages. Here the historical connection of the universities with the church is significant, not primarily because the church has the main function of furthering science, but because the universities, like the church from which they spring, are long-time institutions.

Perhaps our modern scientific foundations may in time achieve a part of the permanency which the universities have already shown, and perhaps if any particular scientific foundation fails to show this permanency, the general institution of scientific foundations and the obligation which is felt to lie on the richer members of the community to establish, maintain, and reestablish them may come to show a part of this permanency. In any case, it is here that the natural responsibility for maintaining a scientific fertility in its highest and deepest levels properly lies.

Once we have established the need for such long-time homes for the intellect, whatever the basis of their administration, we may be able to keep them alive with the aid of secondary criteria which have a valid feedback nature. The long-time usefulness of an idea and the long-time fertility of an originator of ideas are too remote to be judged by direct criteria of performance, but there are certain internal criteria of performance which offer partial indices of some worth. Good scientific work of the sort that is likely to lead to important discoveries shows a texture of consistency, of imagination, and of venturesomeness. This often distinguishes the promising concept from the merely banal suggestion, many generations before this suggestion may find adequate fruition in action.

In speaking of the need for a long-time feedback and of its necessary involvement of faith, I wish to point out that faith itself is subject to a long-time feedback criticism. A faith that has become ossified and that is pursued without any reference to the actual indications that it has performed its function for a considerable period, and is continuing to do so, is an idolatry. Idolatries can be political or economic as well as religious, and there is one political idolatry towards which the ancient Chinese have shown a great understanding.

The notion of the divine right of kings is to be found all over the world, and not least in China, but the Chinese interpretation of this divine right is peculiarly modern. According to Mencius, and indeed to Confucius, rule is from heaven and is carried out under the mandate of heaven. If, however, the welfare of the people suffers and continues to suffer, it is an indication that not only the Emperor, but perhaps the dynasty, has lost the mandate of heaven and deserves to fall. This may be interpreted as an ancient Chinese acceptance of the principle of feedback, even in matters of faith.

The chief difficulty of long-time strategies remains, however; namely, that in adopting such a strategy, we are taking steps for the benefit of our distant descendants, or for those who may stand in their place, rather than for ourselves. We may not only be ignorant of the best strategy for their sake, but we may even be indifferent to their existence, and we certainly cannot hope for such a strategy to produce the clear and indubitable evidence of its effectiveness which we demand for short-time policies. In the matter of corporations and other business firms, this difficulty is peculiarly intimate, for the business firms which have been in existence for more than a century are rare indeed. Those which have been in existence for millennia are practically, if not absolutely, absent. The bank which spends some of its resources for contingencies more than a century away will be regarded as foolish even by the most lenient bank examiner.

Yet if the race is to continue to exist, and I believe there are few of us who are completely indifferent about this matter, some part of our thought and policy must be devoted to long-range undertakings. How, then, can we regulate long-range undertakings in such a way that our policy has a certain amount of confirmation as to its validity?

This is no new problem in society, for the continued existence of every agricultural community has often depended and must depend on the maintenance of irrigation undertakings, and has always depended on the fertility of the soil. The duties of maintaining these long-time assets have been accepted by every society which can term itself civilized in any degree, and they depend ultimately on a certain act of faith as far as the long-run ends themselves are considered. As Tennyson's Old Style Northern Farmer says on his deathbed:

Parson's a bean loikewoise, an' a sittin'
'ere o' my bed.
'The Amoighty's a taakin o' you to issen, may
friend,' a said,
An' a towd ma sins, an' 's toithe were
due, an' I gied it in hond;
I done my duty boy 'um, as I done it boy
the lond.

Compare this with the more worldly view of the Northern Farmer, New Style:

Dosn't thou 'ear my 'erse's legs, as
they canters away?
Proputty, proputty, proputty—that's
what I 'ears 'em say.
Proputty, proputty, proputty—Sam, thou's an
ass for they pains;
Theer's moor sensi i' one o' 'is legs, nor in
all they brains.

As we have seen over and over again, the conservation of the fertility of human thought is as primary an obligation as the conservation of the fertility of the land. Both of these redound to the generations to come, and can only be carried out by one who feels a responsibility, if not to the eternal, at least to the very distant future.

In no sense can such a responsibility be of direct benefit to those on whom it falls, nor even to anyone they know or to anyone who may be considered nearly related to them. Unless society has in it some institution or at least some traditionally accepted mode of behavior pertaining to the very distant future, the long-time care of the future needs of the human race is something which falls equally on everyone, and hence falls on no one.

Merely conventional rules are both too much and not enough to determine these working values of science, but the values themselves are neither trivial nor unrecognizable. They are subject to an intelligent (if restrained) criticism, both from the outside and from others, and within the conscience of the scientist himself.

For this end, however, the scientist must have a conscience and a devotion, and the inner drive which will never permit him to be satisfied with less than the best work which he can perform by his own lights. This sense of mission may be very remote from any formal religion, but the stuff of religion is in it. "The word killeth, but the spirit giveth light."

Patents and Invention: The American Patent System

10

The primitive manner of holding an invention for exploitation is for the inventor to exploit it as a secret, or to hand his secret over for a price to his master or another craftsman. The patent originally came into existence as a method of combating this secrecy. The inventor, who was originally a craftsman in a particular art, agreed to disclose his invention to the public for the sake of the future good of the art, in return for a temporary and limited monopoly granted by the government, and transferable to a possible purchaser. This was, of course, contingent on the existence of a sufficient degree of originality and merit in the invention.

The patent was originally a device for the assurance to the inventor of salable rights in his invention. With the advent of the industrial laboratory, the individual free-lance or shop inventor has been largely supplanted by the man who is hired to invent, at what is often a quite good salary, but who is bound to transfer each invention to his employer for a nominal sum of money. Thus the emphasis of patent protection has passed from the inventor to his employer. The entrepreneur needs the patent system to protect his investment in a new industry by a temporary monopoly. A patent is also a very real protection against being excluded from a certain industrial field by patents that have been obtained by someone else.

Certain defenders of the patent system go so far as to base its desirability solely on the protection that it gives to entrepreneurs, and not at all on the protection that it gives to the individual inventor. It is at least certain that the shop patent system has resulted in many small, insignificant, and vexatious patents, which have cluttered up the art of invention, without contributing meritorious new ideas or new techniques.

The law as to what an invention is should take a real cognizance of what the art of invention is, and of how inventions are actually made. Indeed, in a world which is as dependent on the continued process of invention as ours is, one might expect that the study of the circumstances under which invention takes place would be a primary interest of modern man.

To a certain extent, it is. The law of the Patent Office represents a highly developed code by which property rights in invention can be evaluated, and we have decision after decision handed down by the courts as to what invention is and when it may be presumed to have been made. Yet it is this very interest in invention as an ownable asset which is one of the chief reasons why our knowledge of the process of invention is so defective.

We have a similar situation at the race track. Theoretically, our horse races are a device to ensure the improvement of a breed of horses, and by which this improvement can be tested. Actually the race track is the last place to go to study the genetics of *equus caballus*. In the first place, a very large part of horse racing is done with geldings, animals deliberately rendered sterile and taken out of the game of breeding. In the second place, horseflesh has ceased to correspond in any very close way to the present needs of the human race, and that horseflesh which is still useful for pulling carts, riding range, and other similar socially significant purposes bears a very loose relationship to those elegant counters in a gambling game that are to be seen at Hialeah and Aqueduct.

I will not deny that there is a certain amount of diluted and twisted knowledge of the genetics of the horse to be gathered around our tracks, and still more about our stud farms, but this information primarily concerns the study of the genetics of horse, not as horse, but as a four-legged roulette wheel. These gambling implements have only the slightest relevance even to the cavalry horse, which represents the highly specialized use of the animal as a weapon. In the present days of mechanized cavalry, the whole question of improving the breed of the race horse is remote from any military question. Indeed, it is as remote from any outside practical significance as is the art of breeding Dalmation dogs to run under the old-fashioned horse-drawn fire engines.

This may seem rather far from the question of invention, but the legal aspect of invention and patents has become primarily a set of rules for the larger-scale gambles of modern industry. Patent law and the process of establishing a monopoly in a new turn of art may hold a past significance as a way of developing and rewarding the otherwise anonymous talent to be found in our machine shops. However, the more explicit and conscious supporters of the present patent system have realized that the day of the shop inventor is largely over, and that the process of invention has been changed economically from the sporadic work of the artisan of toil to the function of great research staffs, maintained either by the government or by specific industries. The inventor has been relegated to the function of a race horse in that larger gambling game which constitutes modern industry and modern large-scale undertakings.

Gradually this new point of view has filtered through from the industries to the patent lawyer, and from the patent lawyer to the courts, so that the existing body of doctrine concerning what is invention and what is not invention is about as relevant to the intellectual process of invention as the rules of the Jockey Club are to the real improvement of horses in general.

This use of invention as a gambling token of the game of investment and speculation contributes another motive to those which we have listed for new inventions. These motives have been most varied.

There is a climate of invention that changes from century to century, from decade to decade, and even from year to year. This is of the most varied nature. In one case, the motive for innovation is the discovery of a new technical invention such as the steam engine, or the electric motor, which demands to be evaluated as far as concerns its importance in human society and its social consequences. In another case, the innovation is first felt on the philosophical level, perhaps as a change in the methodology of pure thought.

In still other cases, a particular science like physics begins to outgrow its old frame, and there is a feverish attempt to find a new bottle into which to pour old wine before it all leaks out onto the ground.

Whatever the source of new epoch-making ideas may be and the region in which they first disturb the controls of thought and action, they represent much more than a possible innovation in the field in which they are originally introduced. They spread to other fields by the most diverse and even grotesque channels. Perhaps some individual working in a field as yet unaffected by these new ideas but greatly troubled by its own internal problems happens to be an afficionado of the discipline in which the original mutation occurred. Perhaps the scientist has no such erratic and general interest, but has simply happened to receive his education in some region important to him under the new dispensation rather than the old. Perhaps the trigger of the new thought may be a casual lecture, technical or popular, or even (heaven save the mark!) a story in a book of science fiction.

However the new ferment may spread, intellectual ferments, like biological ones, are infectious. There is, indeed, no way to be touched by a new idea and yet to lapse casually into one's original state of innocence. Thus each generation sees some field which was originally aloof and arcane not only develop in the direction of an improvement of its internal structure and repertorial contents, but also become a source of secondary development both for the other specific intellectual interests of the day and for the general intellectual climate.

To give a short and yet weighty instance, it is still possible to write an un-Freudian novel, but it is quite impossible to write a novel of which we can also be sure that it would have been the same in form and content if the world of Freud had never existed. In all our intellectual and artistic growth, in all the ramifications of our culture, to taste the fruit of the tree of knowledge is an irreversible process.

The fundamental fact of the growth of science and the growth of technique is thus that we are living in a world of changing intellectual climate, and that the spiritual caviar of one generation becomes the curds and whey of the next. This does not mean that the new ideas are appreciated at their first inception by any very large public of progressive thinkers. It does mean that they are there to be appreciated and that there is an infection which is bound to be taken up within a limited time by a sort of chain reaction among the spiritually nonimmune.

In other words, the virus of the new idea is there, and sooner or later it will break out not merely in one focus but in many places. Perhaps indeed it is only those who are exceptionally susceptible who will break out in new thoughts, but this degree of exceptional susceptibility need not be the one-in-a-century grade, or, to modify a figure of speech slightly, when we see a single toadstool push up its head through the floor of the forest, we may be quite sure that under this floor there is a tangle of mycelium and that if we are only sufficiently observant and sufficiently patient, we shall find many other examples of a growth of the same sort.

Intellectual innovation, whether it takes the form of discovery or invention, or the development of new tastes in beauty and new forms of art, is neither absolutely omnipresent nor absolutely sporadic. When an invention such as the telephone is made, if you look through the technical literature, you will find in the most diverse places articles and patent claims that cover other devices which it would not be too fanciful to term telephones.

In those branches of intellectual work in which the reward is purely intellectual and does not take the form of a salable invention, convertible into gold at the proper rate of exchange, it is the mere ordinary stuff of magnanimity to expect that when one comes upon a new idea, one need only search carefully enough through the literature to find the same idea, perhaps in its full perfection or perhaps in a less completely expressed form, again and again.

When, however, one is engaged in the commercial enterprise of selling a patent or exploiting a new invention, the ethics of peace and magnanimity no longer holds, and it is each man for himself and the devil take the hindmost. Under these circumstances, one does not merely notice the mote in the eye of another even though there is a beam in one's own eye, but one carefully cultivates this blemish in others, even where it really does not exist in the first place.

We are all of us familiar with the inventor-manqué, the poor fellow who goes around telling the world that if the lawyers had been a little more fair to him, he would have had a fortune for the invention of the telephone, or the phonograph, or the radio. When this individual becomes, as well he may, a stalking-horse for the nationalism of his own country, we laugh at his claims, as we laugh at the claims of the Russians to have invented everything that has belonged to Edison or Pupin or Steinmetz. Yet those who are familiar with the history of invention do not laugh too loudly, for the gross history of completed inventions which have been fortified by patent law and commercial sale merely represents that minimal part of the iceberg of invention which happens to lie above the water level.

To establish a patent, one must invest what is often a not inconsiderable sum of dollars in legal and technical advice, and if one is to belong to those few who not only have invented but who have made a successful career of invention, one must be able to supplement the original handsome-looking document by a willingness to fight for one's own, which is expensive indeed. In all patent systems, the protection given by a nonadjudicated patent is imperfect, but in our system above all others, a patent is nothing more than a ticket to litigation.

Our flock of young inventors seems to suffer under the delusion that the possession of a patent gives a reasonable presumption of the possession of salable and valuable rights. Nothing could be further from the case. The degree of presumption created by the existence of a patent certificate is slight under the best of circumstances, and carries practically no weight against strong established interests able to procure the best of legal advice, and able to tire out those who weakly possess a good and valuable idea, until they drop it of its own weight.

We have seen an example of the working of our patent system in the moral tale of Heaviside and Pupin. It is hard to see how judges without an engineering training could have been expected to adjudicate a matter of such technical complexity, in a field in which they were totally devoid of training.

I do not for a moment mean to cast doubt on the sincerity of those judges who have upheld the Pupin patents, but I must register a very strong protest against the system by which such a delicate discrimination of right and wrong is left to people with no technical training. It is even more serious that our judges of patent law have no intimate knowledge of the real process of invention and discovery.

I am quite willing to admit that one main object of the law is not theoretical justice but a determination of rights such as the right of ownership, which would preclude later bickering and which will give a decision that is unequivocal and irrevocable, whether or not it can be based on abstract principles of justice. However, when a law is based on a theory of what has actually happened, which does not fit the facts, it is hard or even impossible to adjudicate disputes in an impartial way. The pressure of power, riches, and the ability to use these for the employment of the best legal talent is almost certain to dominate the final decision.

It is an almost universal complaint of patent lawyers that they have to plead before judges who have no training in the technical aspects of the case, and no adequate way of learning what has really happened. It is quite possible that this difficulty belongs to the essence of the situation and that technically trained judges would merely replace one factor leading to unsatisfactory judgments by others of equal seriousness. Nevertheless, a judge without a technical engineering and scientific training must depend on experts to bolster up his judgments, and the experts who come forward in a patent case, as in many other cases, are ex parte witnesses hired by one side or another, who must combine the incompatible functions of expertness and of advocacy.

It has been suggested at times that in addition to the experts testifying for one side or the other in a patent case, there should be still other experts retained by the judge as amici curiae (friends of the court). There is a great deal to be said for this, even though such a course of procedure would have severe disadvantages and limitations of its own.

The fees for public service are highly limited, and do not begin to compare with the rewards that an expert receives for testimony which will establish or reinforce a rich corporation in the possession of valuable rights. Under the circumstances, there is a strong temptation for a great corporation to use the system of retainers for cornering the market in witnesses of top reputation.

Moreover, when the expert witness is asked to express an opinion under conditions where the legal categories have no very close relation to the categories of fact, it is honestly impossible to prove that any opinion may not be sincere, and to establish the fact of perjury. It is my opinion that expert testimony in patent cases, like expert testimony in many others, is shot through with the essence of perjury, at least to the extent that witnesses frequently find themselves in the position of maintaining ex parte opinions which are really not their considered opinion, or which, at least, would not be their considered opinions if they felt themselves free to look at the issues equally from all sides.

However, this perjury of opinion is almost impossible to establish before a court of law to such a point that it constitutes a real risk for the man offering a paid opinion in accordance with his brief. Moreover, an expert witness who lets slip pieces of testimony which weigh against the side that is employing him will find that the very lucrative business of testifying for the strong is closed to him.

So long as judges and witnesses, and even the public at large, remain ignorant of the actual process of discovery, and so long as the greater court consisting of the man in the street does nothing but echo this ignorance, patent procedure will continue to carry the very questionable coloring which it does at the present day. This is true whether the ignorance in question is a real lack of knowledge, or whether it is a voluntary ignorance propped up by the superior financial advantages of the one-sided testimony which makes the expert and the judge alike unwilling to look the real issues in the face.

If I were asked to give a positive opinion of what methods and measures should be adopted to improve our patent law, I must confess that I would be hard put to it. I have spoken about the desirability of including experts as witnesses on behalf of the court rather than on behalf of parties to a patent application or to patent litigation. I certainly think that such a use of witnesses should be encouraged, but I do not believe that this policy would yield anything like a complete solution to the difficulties.

I have a great distrust of the expert as such, and although the public service might draw good men to act as forensic experts, there would be a considerable tendency for a soft job like this to appeal to the conventional and the complacent. Moreover, so long as the real problems of testimony continue to belong to the retained expert of a big corporation, the weaker litigant and the court will have to pick their experts from well down in the apple barrel.

This difficulty seems to me unremovable, even though it can be mitigated. In fact, I would like to see the expert, no matter by whom he is retained, derive his status from the court itself, and act as an officer of the court, with fees to be determined by the court. I do not make a panacea of this, but it would put the whole issue of expert testimony from the beginning on a more dignified basis than that on which we find it at present.

The peculiar vices of the hired expert retained for one side or another, which beset the patent system in this country, are not unrelated to the relatively low evidential value of a United States patent. As compared with German patents, at least in the normal periods of German life before World War I, and to some extent between the world wars, American patents were inexpensive and easy to obtain, but they were of low evidential value.

The German Patent Office, like most European patent offices, insisted on a very exhaustive search by expert patent examiners before a patent could be granted, and the expenses of this search were paid by the applicant. There were other, more severe conditions attached to German patents, as compared with American patents, as, for example, the unwillingness of the Germans to establish paper patents, which was expressed in their insistence that a patent be exploited industrially within a certain limited time or else lose its force.

Thus a German patent was something to be taken very seriously indeed, and it had a high presumption of validity before the German courts. Conversely, the American patent, and especially that of recent years, was granted rather lightly and had very little presumptive value as evidence that a patentable invention really exists. The American system, which has indeed somewhat declined, of building up a ring fence of paper patents for the purpose of protecting the interests of a company in directions in which it might possibly want to work, and of embarrassing competitors without any real intention of exploiting the patents, has contributed to the great number of patents taken out in the United States, and has further lessened the evidential value of each patent.

The effect has been that an American patent has not been a safe basis for the establishment of a business until it has been through a certain amount of litigation. The result is to transfer the real authority in patents from the examiners, as administrative officials, to the judiciary. The examiners in Germany are essentially experts who cannot act ex parte, and the large degree of evidential authority accorded to their judgments has an effect very similar to that which expert witnesses acting as friends of the court would have in patent cases.

The system of highly expert examiners whose judgments have a large evidential value puts the primary authority in patent matters where it belongs, and that is with people with a specific engineering and patent training. Our average federal judge knows very little about patents, and this little he has learned in a rather fortuitous manner. It therefore seems to me that the German system is fundamentally a better one, and is more able to take cognizance of changes in the art than is ours. A highly trained patent examiner will become aware of a change in the art at its source, while our judges in patent matters are not primarily patent judges, and have too little experience with such cases to build up a really personal acquaintance with the present state of the arts.

I am suggesting something like the European patent system as a system which is better intrinsically. Whether it is better in practice, and whether the various pressures to which patent authorities at all levels are subject would not sooner or later bring them to a state not much better than that of the patent judges, is something that I am not prepared to say. This remains for somebody with much more thorough knowledge to decide, and this knowledge must cover not only the theory but the practice of both our judicial and our administrative systems.

The question of the difference between an invention and a law of nature needs new laws, but even more it needs new thinking among patent authorities. A piece of work which is patchy and incomplete may seem to be an invention, whereas a more thorough study of it might reveal its essence to be a law of nature. Thus it may not pay an inventor to understand his invention too well, and he may even lose rights in it as a limited invention by being fully aware of a larger law of nature of which it is a part. This is to my mind an intolerable state of affairs: that sloppiness and dullness should be rewarded, and that an intelligent understanding should be penalized.

As to the discoverer of a law of nature in the strict sense, it is manifestly impossible and inequitable to give him a long-time control in something which will affect all science from his time on. No man deserves so much power, and no man will benefit by so much power. Nevertheless, it is conceivable to me that an enlightened policy of governmental rewards might be adopted for those who have done fundamental work of discovery.

In my next chapter, I shall discuss this in greater detail, but whether these rewards should take the form of prizes or of premiums, not so definitely limited in number as are prizes, or even, in extreme cases, of pensions, I do not know, but I do not consider it impossible that some system may be worked out which may add to the desirability or, at any rate, to the security of a life career in science.

All of these improvements of technique, however, fail to touch the real nub of the problem. This is that lawyers and judges in general, and those who deal with patent rights and other intellectual property in particular, should be better educated not only in technical details but also as to the actual processes which constitute invention and discovery. I doubt if this is a matter which is at all readily susceptible to any improvement by enactment or by change of procedure. It must come as a phase of a better and wider understanding of invention and discovery on the part of the community itself. If I shall have failed to contribute to this in some small degree by the present book, I shall indeed have failed in my purpose.

Goals and Problems

11

The years since World War II have seen an enormous change in all our policies, and not least in our scientific policies. It has been a period of great projects, of the partial eclipse of the individual, and of the emergence of a mass attack as a normal method of scientific work. It has seen the universities and other classically scientific organizations and organizations of learning facing a tremendous inflation which has decimated their resources and has made it necessary to address new appeals to the public.

This public has become emergency-minded, and therefore prepared to sacrifice long-time ends and ideals of scholarship to the combating of the dangers of the moment. These dangers include a world struggle for supremacy between the United States and Soviet Russia. Like the United States, Russia is an aggressive nation with a philosophy conditioned not so much upon stability as upon continued progress towards a goal not yet attained.

However, our struggle with Soviet Russia and the Soviet complex of nations, which has already dwarfed the indecisive medieval struggle between Christianity and Islam, is only one of those which the present generation must face. Militant capitalism—and not all capitalism is militant—is a creed which represents as much of a departure from much of our nineteenth-century traditions as does Communism, particularly when it is combined with a pseudo-egalitarianism which claims not only that one man is as good as another, in his moral rights and duties, but that he is interchangeable with another in his abilities.

The present disputes and controversies in which we have engaged have lasted over decades. However chronic they may prove to be, they have attracted our attention and concern in an acute form, which I do not think has yet settled down to a long-time equilibrium. In this phase as short-time conflicts, they have drawn our attention away from our long-time problems. They have even made large sectors of our public opinion scarcely aware that the long-time problems exist. I have asserted, and I think that to a certain degree I have shown, that the long-time problems must be faced by long-time social feedbacks, or, if you prefer, checks and balances, if we are not to sell the future of humanity short.

I have already pointed out that in an athletic contest which demands vigorous exertion and even maximum exertion, of short duration, we can ignore with impunity the long-time needs of the athlete. The sprint runner may burst his heart if he is out of condition, but he will not die of starvation because of the immediate demands of the contest. He will not even die of oxygen starvation, for there is a considerable amount of oxygen already held in combination by the hemoglobin of his blood and of his tissues. He cannot exhaust this reserve in a short sprint.

But not all athletic contests are of short duration. No marathon racer can continue to run if his rate of consumption of oxygen is greater than the rate of intake of oxygen into his blood. He must reach a quasi-equilibrium in his needs which will last for the whole period of the race. The legend of Phidippides going mad and dying at the end of his run from Marathon to Athens is exactly what one must expect on such an occasion. Similarly, a nation, or the human race, which must face a protracted period of emergency, can by no means ignore the continual satisfaction of its long-time needs.

Now, a great war does not usually last more than some five years, and even the cold war, which belongs to the times in which we live, has not yet endured more than a decade. When, however, our President tells us that the emergency of the cold war and the threat of a new hot war will be with us for some forty years, we cannot be content with meeting this threat by measures which depend solely on the consumption of our stored resources.

Forty years from now, our leading inventions are likely to be based on scientific ideas not yet even contemplated. The relative rate at which Russia and ourselves store up reserves of fundamental science is more likely to be a fair measure of our ability to meet Russian competition forty years from now than is our present level of technical achievement on the basis of the scientific concepts which we have already mastered. There is a certain and very real evidence that Russia may have been devoting more thought to the long-time problem of maintaining the level of science than we have. If we are not careful, this may offset all our technical know-how and all the concentrated large-scale effort which we have shown ourselves capable of making.

Russia may be our chief antagonist or potential antagonist for many years, perhaps even for more than the forty of which President Eisenhower speaks. However, our long-time chief antagonist will not be Russia, but is to be sought for among the continuing threats of hunger, thirst, ignorance, overpopulation, and perhaps the new dangers of the poisoning of the world in which we live by the radioactive by-products of an atomic age. Thus we must go into training for a marathon race and not a sprint. We cannot win in this marathon race unless we have that sense of the future which is based on a sense of history.

This book, then, is a plea that even in this present difficult and confusing period, we must take a long view and conserve long-time values. I will not say that our present effort is not enough, for it is nearly a maximum effort. It is certainly not, in its entirety, the right sort of effort for the problems which will continue to confuse us. We cannot survive our long prospective ordeal by virtue of adrenalin alone.

Let me sum up certain considerations already explicit or implicit in the earlier chapters of this book. We have already seen that one of the basic observations of the history of invention and discovery is that the really great steps in invention are, at least in many cases, nothing but the embodiments of changes in the intellectual climate, which often antedate their industrial use by several decades. In this preindustrial stage, the new ideas might or might not have been introduced, and the lack of one or two individuals in the chain of thought leading up to them, while it might not have made the new developments forever impossible, could easily have delayed them for over a generation.

On the other hand, the later stages of invention, which often result from the building up of general scientific education, and of thought in many separate fields, to the point where the new ideas can be appreciated, have a very strong tendency to be a repeated and not a sporadic phenomenon, and to take place nearly at the same time in many fields of work, among many individuals, and in many countries. What I wish to discuss in this final chapter is how such a multiform human activity can best be allowed to develop in its fullness and what measures suggest themselves for fostering it and its various diverse stages.

By the time that the second period of invention is well along, and the inventors are working with ideas which thrust themselves upon them from many different quarters, it slowly begins to be possible to evaluate the repercussion they will have on human affairs generally, and on manufacturing and engineering technique in particular. The questions oflegal ownership and protection still need the expert services of the patent lawyer. Apart from this patent phase of the development of invention, it is possible to envisage a program of development and exploitation which, if it involves elements of risk, on the whole involves a calculable risk.

With such a calculable risk in view, it may easily pay a commercial organization, or a pseudo-commercial organization of the nature now to be found in many government departments, to put onto the idea a large and subdivided group of engineers and scientists, and even to assign each man in that group a specific task which he is to accomplish. If this development were completely without risk, it would probably also be without reward. The picture I am giving contemplates a certain risk, and bears a not-too-remote resemblance to a commercial undertaking of the ordinary sort in which new ideas do not necessarily play an essential role.

The mass-attack, piece-work concept of the development, research, and engineering functions of large organizations has captured both the public fancy and the fancy of our entrepreneurs and government officials. It has captured this fancy almost to the exclusion of any other concept of discovery or invention. I have said and I here say again that it is a concept which cannot be extended to the complete history of discovery, and which is particularly inapplicable to the first and most vital steps towards invention.

In their early stages, invention and discovery are not calculable risks. In the first place, the really inventive mind must take chances. A long-time record of no false starts probably does not mean infallibility on the part of the inventor, but merely that he has not been willing to push his ideas as far as they merit. The baseball player without a record of errors is the one who does not go after the balls that he might barely pull down, but allows them to damage the record of the other fellow.

In the second place, the really big finds in invention and discovery take a long period to mature. Very often they cannot be put to any economic use until other ideas as yet undiscovered have rendered practical what was once a long shot. There is no way of keeping an invention on ice for a long period and still exploiting it commercially. To keep a discovery secret means an unwilling or willing denial of the idea advantages of the discovery to other people. After a very limited period—limited as far as really fundamental inventions are concerned—to publish it means to dedicate it to the public. As we have seen in the case of Heaviside, this means to deny exclusive commercial rights to any one person. Thus the probability that a really fundamental invention will accrue to the commercial advantage of the person or group who has invented it is very small indeed. It is simply not worthwhile to invest a cent in it, as there is no process by which this bread cast upon the waters will return to the same person who has set it afloat.

Our economic setup takes cognizance of large calculated risks, but it does not take cognizance of small uncalculated or incalculable risks. This is consistent with the fact that it is to the advantage of the community that the community as a whole, which means simply some persons in it, should take these risks, but they are outside the normal specific profit and return setup. Who is to carry a new idea from conception to parturition?

In our community as it stands, we have scientific institutions, such as the universities and a few of the foundations, which are run on a nonprofit basis. What is more, we have a few, but a very important few, individuals in the community who are only moderately sensitive to the profit motive. When the important functions of these institutions and these individuals are taken over by industry, there results a very subtle but significant distortion of their purpose.

Certain industries and certain companies have, in fact, maintained highly individualistic scientists and highly individualistic sciences on their payrolls. To some extent, this is because the same individuals and sciences that contribute to the community at large and to the environment that a particular firm finds profitable may be kept on tap for more specific problems of development. In this double use of them, there is also the risk that they will find themselves swamped by these problems and cease to contribute to the science on a larger level.

Some individuals, and these sciences, may also make a certain secondary contribution to industry in the form of good name and advertising. Charles Steinmetz's association with the General Electric Company resulted in very important and immediately usable inventions, but it also resulted in conveying an impression to the public that something important to science was happening in the General Electric Company. This was not without its value to their advertising men and to their good name.

The case of Steinmetz was, however, a very special one. In addition to real top ability, his deformity, and the individuality of character which was not altogether dissociated from this, made him picturesque. But a Steinmetz is a windfall, and those companies which have tried to make creative personalities of a less picturesque nature subservient to their advertising needs have had far from uniform success in the matter.

An industry may shelter one or two men of this sort, but it cannot, in fact, shelter a sufficiently large and typical body of them to assure itself and the community as a whole of the continual replacement of ideas which is necessary to prevent the well of science from ultimately running dry.

There are other difficulties in the way of the commercial firm when it comes to the employment of the individualistic scientist, more especially when he is employed in the same building as more conventional engineers or development men. It is of the essence of the individualistic scientist that he must take his reward for being what he is rather in freedom than in money. If he is not that sort of person to begin with, he will soon find the development laboratory and even the sales force a more congenial point of vantage from which to follow his main interests.

Yet this freedom which he has will also be a thorn in the side of the more conventional scientists who work in the same building, and even more a thorn in the side of the purely industrial employees. Unless he is a very eminent man, the unconventional scientist will be disliked. Even the eminent man only arrives at his position by going through stages in which he is not eminent.

Thus, for general good will and good feeling, it is to the advantage of the community that the worker in the more fundamental research should have a home in which he is congenial and is wanted, and this home should often be an institution in which his love for freedom and the respect which he demands from his individual associates are not abnormal favors granted to him and to no one else, but a part of the natural environment in which he works.

Combine this limitation on his usefulness with the real doubt as to whether his services to humanity will accrue in particular to his employer rather than to some very different person who finds out what they are at a much later stage, and it will become clear why the commercial laboratory can never become the exclusive home of science.

The present age favors the big laboratory and the man who is sensitive to the profit motive. Moreover, it favors the man who will meet his industrial colleagues at the country club or over the bridge table. The real scientist of the first rank is by the nature of his own activity too busy to care much for either money or the ordinary signs of prosperity with which we have been made so well acquainted by Thorstein Veblen.

It is true that the college professor must fight the peculiar conservatism of the college and the peculiar sort of social acceptance on which it insists, but although these restrictions on his freedom are real, they are certainly no larger than those restrictions which bind the industrial employee. The man who is too much of an individual for the college laboratory will almost always be too much of an individual for the industrial laboratory.

There are other ways of killing a cat than smothering it in butter, but, after all, smothering it in butter is a very effective way of getting rid of it. There are other ways of destroying the intellectual fertility of the scientist than giving him a million dollars to work with and then expecting proportionate financial results, but that is also a good way of putting him out of action. To the able and conscientious man, the control of a large sum of money is always a burden, unless it has a tremendously strong motive behind it. If I have nothing but my salary, I do not feel particularly humiliated if my new idea cracks up and I go for a year or two without a success. If I were employed at the sort of salary which until recently was reserved for royalty and the heads of companies, I would very soon begin to question myself concerning my own sterility. It might well be that others would question me too, including my employer. At any rate, it would be difficult for me personally to accept a fabulous salary without forcing on myself the demand for fabulous work.

The free-lance does not fit easily into any particular niche of a purely capitalist system. To a certain extent, this has nothing to do with capitalism, for he would not fit any better into a large-scale Communist or socialist organization, which is compelled by industrial tradition or by state discipline to budget the unbudgetable, whether in money or in reserve national effort. The "megabuck" scientist, the man who plays at a million dollars a throw, is equally the slave of his position, whether the money comes from a private industry or a state monopoly.

I am making this book a plea for the effective scientist who nevertheless wishes to lead the life of the small man and who, in fact, cannot fulfill his function unless he does lead such a life. I do not say that the capitalist system or Big Business or the Big State is universally invalid. I simply say that they do not represent a complete appraisal of the social needs of the community in the matter of invention and discovery. They all of them represent a certain localization of scientific activity and of scientific responsibility that is not necessarily for the good of the people, the community at large, or even the complete and long-time good of Capitalism or Statism itself.

The profit motive may be important, but it must be supplemented by other motives. The community must cultivate a group which is neither subservient to the profit motive in the external community nor internally governed by this motive. We need the free-lance in the arts and in the sciences, and the free-lance finds his natural home in institutions which are primarily designed to give him his proper scope. For the really vital stages of invention and discovery, the small laboratory and the well-organized foundation, such as the Guggenheim Foundation, which is tailored to the needs of the individual scientist, occupy an indispensable place.

The integrity of discovery and the possibility of a decent career for the discoverer and the inventor must be conserved by methods which are not confined to giving the inventor, his employer, or his heirs or assigns permanent property rights in his thought. This is especially true if we consider, as many people do at present, that the rights of freedom of initiative and of personal ownership contain the implication that the proprietor of any chattel or right has an implicit license to waste, to destroy, or to suppress what is his.

Over history at large, as Professor Karl Deutsch has pointed out to me, there are relatively few epochs in which one's property and long-time assets have been regarded as absolute, and ownership has imparted the right to waste and to destroy the earned fruits of the land, as well as to conserve and to enjoy them. One of these periods was that of imperial Rome and another that of the colonial expansion, beginning with the time of the Spanish conquests and only now drawing to its close. Professor Deutsch has made it clear to me that these periods of unlimited property in land have also been periods of chattel slavery, that is, of unlimited property in human beings.

Most periods of slavery, and in particular that of the serfdom of the Middle Ages, have given the slave owner somewhat limited rights in his slaves. Serfdom attached the serf to the land, and assimilated him to the land as far as his ownership went. At least in theory, a lord had no more right to deal wastefully with the bodies of his serfs than with the soil of his estate. What further prevented him from owning his serfs or his land as his absolute property was that he held both of them with a certain sharp responsibility to the rights of his overlord.

The Roman ergastulum and the slave camps of the Southern plantation; the Mexican silver mines of the sixteenth century and the twentieth-century diamond mines of South Africa: all these belong to a tradition harsher and more wasteful than that of the feudal system.

Behind what I may call the plantation attitude, which permits one to work one's slaves to death and to grab the fruits of the fertility of the soil without maintaining the soil for future generations, lies a direct denial of the attitude of Albert Schweitzer: namely, respect for life. To treat human beings as consumable property assets is, of course, a lack of respect for life; but so it is to chop down a forest without replacing it, to squander the soil, or to render sterile the soil of the human intellect. It is because Schweitzer has seen the degradation which is common to all these, and has denounced it in no uncertain terms, that he has achieved the rank of a modern prophet.

Thus there can be no genuine ownership of a really fundamental idea, but only a stewardship of such an idea in trust for the community. How can such a stewardship best be facilitated by measures which are fair both to the community and to the creative intellect? Without hoping to settle great matters of policy in one or two sentences, let me suggest one or two factors which may be significant.

In order for invention to fulfill its social function it must be recognized. It is not enough that somebody discover a new idea, nor is it even enough that this new idea be conserved in archives accessible to everyone. Information is not merely something which concerns what is said, but rather the proportion of what is said to what might be said.

The greatest book in the world may exist in a library. If there is no system for directing the attention of the reader to that book to the exclusion of other books, it is of little value, and it is of proportionately less value the larger the library is. In other words, the full merit of an intellectual discovery is diminished in the absence of a machinery for the recognition of this discovery.

The world must be full of mute, inglorious Miltons. They cannot perform the function of a Milton so long as they are mute and inglorious, or even so long as no one listens to their voices precisely because they are inglorious.

It is for this reason, much more than for any titillation which recognition gives to the scholar, that it is important that inventive minds be selected from the ruck of less inventive minds by some act of recognition. No man is a real intellectual of the first rank who does not have a pretty good idea of the value or lack of value of his work. False modesty is not one of the major virtues.

Conversely, a gem of purest ray serene is of very little value in the deep unfathomed caves of Ocean. It is not for the creative scientist himself that his recognition is primarily important: it is for the other scientists and for the people who are going to use his work. The principal valid point of a system of rewards and recognitions for the scientist is that it helps the direction of search of the learner and of the man who wishes to use scientific work.

It is neither desirable nor possible to give the discoverer of a brand new scientific idea the full monopoly of the commercial returns of his idea for a period of any length. On the one hand, there are ideas of such consequence that no man should have the returns which an effective monopoly of this kind would bring. I am saying this, not merely for the sake of those who might be subjected to the scientist by this authority, but quite as much in the interest of the scientist himself. The maintenance of the position of being a Very Important Person, even if we assume the greatest internal modesty on the part of such a person, is a burden and an isolation. This is not least so in the case of the scholar, a man who should be up and doing something new.

On the other hand, the value of a piece of scientific work only appears to the full with its further application by many minds and with its free communication to other minds. Here any secrecy or any rights of property possession will naturally have the effect of making people shy off from a preempted field of work. In so doing, they will actually delay the intrinsic consequences of a discovery. As a secondary matter, they will delay the true and full recognition which the discoverer should have through the mass of work dependent on his own.

Since a purely property reward is not in order, there should be, in my opinion, some other system of rewards for scientific work of only remote and contingent practical application. This system should be directed towards the original (not the derivative) selection of the really creative individuals. It should aim at a certain facilitation of their work by allowing them to live in a satisfactory environment, which need in no wise be proportioned to the ultimate commercial value of their discoveries. Here I class discoveries and inventions together, because no exclusive property rights come into consideration. Discovery is as great a contribution to the future of knowledge and technique as is invention.

What forms such rewards should take, and by whom they should be administered, I am not prepared to say dogmatically. There are several alternative possibilities. Besides the prize, which represents an award limited in numbers and not adaptable to the changed level of production that is likely to accompany a universally increased scientific activity, I can think of the premium, in which the award depends more on the absolute merit of the intellectual work accomplished than on its relative merit in a limited competition. I can also think of the pension, which might either be awarded in such a manner as to give an immediate improvement in the scale of living of the beneficiary, or so as to cancel his fear of an indigent old age. I forbear to recommend one of these courses more than another, or to pronounce upon the agencies which should be entrusted with this important social responsibility.

At any rate, whatever benefits are awarded for scientific creation should have the good of the community as their purpose even more than the good of the individual. As such, they should be contingent on a full and free publication of the new ideas of the discoverer. The truth can make us free only when it is a freely obtainable truth.

Index